P9-CDE-863

A TALE OF SEVEN ELEMENTS

A Tale of Seven Elements

Eric Scerri

OXFORD
UNIVERSITY PRESS

Oxford University Press is a department of the University of Oxford.
It furthers the University's objective of excellence in research, scholarship,
and education by publishing worldwide.

Oxford New York
Auckland Cape Town Dar es Salaam Hong Kong Karachi
Kuala Lumpur Madrid Melbourne Mexico City Nairobi
New Delhi Shanghai Taipei Toronto

With offices in
Argentina Austria Brazil Chile Czech Republic France Greece
Guatemala Hungary Italy Japan Poland Portugal Singapore
South Korea Switzerland Thailand Turkey Ukraine Vietnam

Published in the United States of America by
Oxford University Press
198 Madison Avenue, New York, NY 10016

A copy of this book's Cataloging-in-Publication Data
is on file with the Library of Congress.
ISBN 978-0-19-539131-2

9 8 7 6 5 4 3 2 1
Printed in the United States of America
on acid-free paper

Dedicated to my mother, Ines Scerri,
on the occasion of her 90th birthday
and to my wife Elisa

Also dedicated to the centenary of
Henry Mosley's landmark 1913
article on an x-ray method for
ordering the elements

CONTENTS

PREFACE

The first thing to be said about *A Tale of Seven Elements* is that it is wonderfully rich and full, imparting a huge range of knowledge not only on the properties of each of these elements but on the nature of science, the meaning of discovery, and how these are deeply entwined in their social and political context.

In his earlier book *The Periodic Table*, Eric Scerri concentrated on the history and philosophy of periodic systems and the many forms these have taken since Mendeleev's original table. It was evident to Mendeleev that there were gaps or holes in the Periodic Table, and he boldly predicted the discovery of several as-yet-undiscovered elements to fill these gaps. But it was only in 1913, with Moseley's demonstration that the elements had integral atomic numbers that one could confidently say that, of the 92 elements up to uranium, just seven were missing: those with the atomic numbers of 43, 61, 72, 75, 85, 87, and 91.

Scerri expands here on the stories of these elements, which were painstakingly isolated between the First and Second World Wars. These seven elements—technetium, promethium, hafnium, rhenium, astatine, francium and protactinium—are nearly all tantalizingly elusive and difficult to isolate. Four of them are intensely radioactive and exist in only trace amounts, if at all, in the Earth's crust. Their discoveries involved intricate stories of epic labors, inspired detective work, scientific passions, collaborations, competitions, and hopes repeatedly raised and dashed.

Scerri is particularly interested in the bitter and protracted disputes over priority that often arose with these seven elusive elements, and how partisanship and national pride, intensified by the demands of war, inflamed these disputes. What constitutes "discovery"? How do we define "priority"? With so many researchers looking for the same few elements, much is left to chance, to the lucky hunch, to national rivalries, to personal ambition.

Scerri's vivid storytelling, and the letters and journals he quotes, allow us to see chemistry, and science generally, as an essentially historical enterprise—a human adventure that shows the best, and sometimes the worst, of human nature. As a boy I read Mary Elvira Weeks' classic *Discovery of the Elements* with great delight. Now, seventy years later, I get the same sort of delight from *A Tale of Seven Elements,* and I think that it, too, will become a classic.

Oliver Sacks, March, 2013.

AN A–Z OF ACKNOWLEDGMENTS

Jean Pierre Adloff, Roy Alexander, Rolando Alfaro, Santiago Alvarez, Maria Victoria Alzate Cano, Peter Atkins, Jochen Autschbach Eugene Babaev, Philip Ball, Henry Bent, David Black, Roderick Black, Gisela Boeck, Boy Boer, Jan Boeyens, Danial Bonchev, Teresa Bondora, Christianne Bonelle, Marie-Anne Bouchiat, David Bradley, Shawn Burdette, Albert Burgstahler, Sam Cannarozzi, Stuart Cantrill, Davide Castelvecchi, Alan Chalmers, José Antonio Chamizo, Hasok Chang, Oliver Choroba, Stephen Contakes, Dale Corson, Kevin De Berg, Robert DeKosky, Paul Depovere, Pilar Gonzalez Duarte, Fernando Dufour, Christoph Düllmann, Joseph Earley, Robert Eicher, George Ellis, John Emsley, Sibil Erduran, Grant Fisher, Marco Fotani, Jan Fransaer, Gernot Frenking, Andoni Garritz, Carmen Giunta, Theo Gray, Brian Gregory, Alan Gross, Julio Gutierrez, Henry Griffin, Fathi Habashi, Steve Hardinger, Dean Harman, Rom Harré, Ray Hefferlin, Ned Heindel, Robin Hendry, Hinne Hettema, Birger Hjørland, Peter Hodder, Merce Izquierdo, Knud Jensen, Masanori Kaji, Rick Kaner, Paul Karol, Gary Katz, George Kauffman, Sam Kean, Helge Kragh, Werner Kutzelnigg, Martin Labarca, Lawrence Lavelle, Mike Laing, Mark Leach, Jeffrey Leigh, Jean-Pierre Llored, Olimpia Lombardi, Steve Lyons, Robert Luft, John Luton, Charles McCaw, Sharon McGrayne, Lee

McIntyre, Hisao Mabuchi, James Marshall, Cherif Matta, Michael Matthews, Seymour Mauskopf, P. K. Mohapatra, Susana Moreno, Peter Morris, Clarence Murphy, Dimitry Mustafin, Fabienne Myers, Paul Needham, Peter Nelson, Octavio Novaro, Yuri Oganessian, David O'Hagan, Luis Orozco, Rubén Darío Osorio Giraldo, Richard Pagni, Sergio Palazzi, Martyn Poliakoff, Pekka Pyykkö, Guillermo Restrepo, David Robertson, James Rota, Oliver Sacks, Jeffrey Seaman, Siegfried Schindler, Chris Schmidt, Joachim Schummer, Eugen Schwarz, Peter Schwerdtfeger, Andrea Sella, Bassam Shakashiri, Sylvain Smadja, Lee Sobotka, Hristina Stefani, Philip Stewart, Keith Taber, Jess Tauber, Brett Thornton, Pieter Thyssen, Valery Tsimmermann, Brigitte Van Tigglen, Jose Luis Villaveces, Jacob Wacker, John Wasson, Frank Weinhold, Michael Weisberg, Max Whitby, Mark Winter, Donny Wise, Kenishi Yoshihara, Roberto Zingales,

A special thanks to my editor at OUP, Jeremy Lewis as well as all editorial staff.

INTRODUCTION

Over just the past four or five years, the periodic table has become far more than simply a scientific icon. It is fair to say that it has captured the public imagination like never before, in addition to becoming an icon of style and an icon of design used to market all kinds of products. Part of this explosion of interest has been spawned by the growth of the Internet. For example, there are now hundreds of YouTube videos featuring attempts to explain the history of the periodic table or Tom Lehrer's song of the elements sung by anything from two-year-old children to established movie actors. There have been several popular books on the subject as well as iPhone and iPad applications.[1] Every day or so, it seems, sees the invention of a new periodic table of some particular domain. Two of my own particular favorites are the periodic table of guitarists and of jazz music.[2] Even fine artists and advertisers have embraced the periodic table, sometimes transforming it in ways that are almost unrecognizable. In addition, there are websites dedicated to the variety of scientific periodic tables that are continually being devised by professionals and amateurs alike. We should also not forget what my friend and fellow periodic table author, John Emsley, has written about the table:

> As long as chemistry is studied there will be a periodic table. And even if someday we communicate with another part of the universe, we can be sure that one thing that both cultures will have in common is an ordered system of the elements that will be instantly recognizable by both intelligent life forms.[3]

The present book is aimed at readers who are interested in digging a little deeper into the science of the elements and the periodic table. The writing of the book has been motivated by a number of factors. It came about as a result of wanting to write a follow-up book to my book on the periodic table.[4] One day while reading, I rediscovered an interesting paper by Vladimir Karpenko on what he calls "spurious elements," meaning elements that were announced but later turned out to be incorrect for a variety of reasons.[5] This article reminded me of an ongoing controversy concerning element number 43 and the fact that Professor Fathi Habashi, a metallurgist based in Quebec City, had written to me to point out that I had made an error in my book when briefly discussing the discovery of this element. Professor Habashi sent me a copy of a letter he had published in the *Journal of Chemical Education*.[6] In it he had pointed out that the work of Van Assche and Armstrong, in their attempt to rehabilitate the claim made by the Noddacks in 1925 for having isolated element 43, was in fact untenable.

Within a couple of days, it suddenly came to me. I would write a book to examine this case as one of the seven elements, which represented the missing gaps in the periodic table after Moseley had established his elegant method for "counting" the elements. Another of these seven elements is hafnium (# 72), on which I had written a couple of papers and whose isolation was also a controversial issue.

In addition, another interesting "element" emerged, if you will excuse the pun. I realized that of the seven elements in question, all of them rather exotic (technetium, promethium, hafnium, rhenium, astatine, francium, and protactinium), three or possibly four of them had been first isolated by women (Meitner, Noddack, Peyer). And if I were to begin my story a few years earlier, I would need to include the even more famous Marie Curie, who quite definitely isolated two elements, polonium and radium. Although it is something of a truism that women were and still are highly unrepresented in the sciences, the discovery of the elements has been one area where they have been rather influential, even if they have not always received their due credit.[7] But the idea of the seven elements that are rather nicely delineated from the others, as I will explain, made me decide to go with the idea of "A tale of seven elements."

I should expand a little further what I mean by the seven missing gaps in the periodic table. I am referring to the fact that after Moseley had developed his method, which will be discussed in due course, it became clear that there were seven elements yet to be isolated among the ninety-two naturally occurring elements from hydrogen (#1) to uranium (#92).[8] Now again, this apparent simplicity is somewhat spoiled by the fact that, as it turned out, some of these seven elements were first isolated from natural sources following their being artificially created, but this raises several more issues that are best left to later sections of this book. The fact remains that five of these seven elements are radioactive, the two exceptions being hafnium and rhenium, the second and third of them to be isolated.[9]

It could be said that there were many more than seven gaps in the periodic table since a remarkable further twenty-six elements, lying beyond uranium, have been discovered and have taken their places in the periodic table.[10] Indeed, one of these elements, neptunium, was synthesized before the seven gaps within the elements 1–92 had been filled. These synthetic elements had not even been envisaged at the time when Moseley devised his method. Their discovery, or more accurately their creation, will be discussed in the last chapter of the book.

So given this somewhat arbitrary decision to begin with the above-named seven elements, I am nearly ready to begin the tale. But first I should say something about scientific stories. There is now a growing literature on the importance of stories and narrative in science as well as in science education.[11] Given my long-term interest in the history of science, and my growing interest in science stories, I knew I had arrived at an interesting topic for a book. I can only hope that you will agree.

What remained was the question of the order in which to consider these elements. Should I simply follow an alphabetic order as done in the excellent books on the elements by Emsley and by Stwertka?[12] Should I take the more naturalistic approach of following the order of increasing atomic number, especially given the importance of atomic number in actually identifying the remaining seven elements in the periodic table? Or, given the historical dimension of this project, should I follow the chronological order of the seven elements' isolation?

After much thought, I decided to follow a historical order of the discovery of the seven elements, starting with protactinium in 1917 and ending with promethium in 1945 (fig. 0.1).

This period, therefore, roughly spans the two world wars. In some cases the state of war that existed had a very direct influence on attempts to isolate some of the seven elements. This is true of protactinium, for example, the first of them to be isolated. In 1916 its discoverer, Lise Meitner, writes of the problems she encounters in purchasing even the most basic items of laboratory equipment:

> Dear Herr Hahn!
>
> The pitchblende experiment is of course important and interesting but you must not be angry with me if I cannot do it right now...I have ordered the vessels for our actinium experiments, will get them in a few days and will begin right away...Be well and

Meitner	Hevesy	Noddacks	Segrè	Perey	Segrè	Marinsky
91	72	75	43	87	85	61
Pa	Hf	Re	Tc	Fr	At	Pm
1917	1923	1925	1937	1939	1940	1945

FIGURE 0.1 The seven elements in chronological order of their official discovery, including names of most senior discoverer and atomic numbers.

> please don't be angry about delays with the pitchblende. Believe me, it is not because of lack of will, but because of lack of time. I can't very well do as much work alone as the three of us did together. Yesterday I bought 3 meters of rubber tubing for 22M!! I got quite a shock when I saw the bill.[13]

In the case of the second war, the discovery of nuclear fission, also by Meitner and her associates, quickly led to the development of the nuclear weapons program and bringing the war to a close. Nuclear research also led to the development of particle accelerators. And it was by using a particle accelerator that some of the remaining seven elements were first identified, such as promethium in 1945, as World War II was drawing to an end. It goes without saying that scientific discoveries occur within a social and political context and the discovery of the seven elements is no exception to this trend. Moreover, the twenty-six elements that lie beyond uranium have all been

H																	He
Li	Be											B	C	N	O	F	Ne
Na	Mg											Al	Si	P	S	Cl	Ar
K	Ca	Sc	Ti	V	Cr	Mn	Fe	Co	Ni	Cu	Zn	Ga	Ge	As	Se	Br	Kr
Rb	Sr	Y	Zr	Nb	Mo	**Tc**	Ru	Rh	Pd	Ag	Cd	In	Sn	Sb	Te	I	Xe
Cs	Ba	Lu	**Hf**	Ta	W	**Re**	Os	Ir	Pt	Au	Hg	Tl	Pb	Bi	Po	**At**	Rn
Fr	Ra	Lr	Rf	Db	Sg	Bh	Hs	Mt	Ds	Rg	Cn		Fl		Lv		

La	Ce	Pr	Nd	**Pm**	Sm	Eu	Gd	Tb	Dy	Ho	Er	Tm	Yb
Ac	Th	**Pa**	U	Np	Pu	Am	Cm	Bk	Cf	Es	Fm	Md	No

FIGURE 0.2 Positions of the seven elements (shown in bold letters) on a medium-long form periodic table. The elements in this and many following tables include all the elements known today.

synthesized using accelerator technology of one form or another, as will be described.

This book is primarily about the final seven elements to be discovered among the first ninety-two elements that I propose to call the "infra-uranium elements" by contrast to those beyond uranium or the transuranium elements.[14] These elements appear in all parts of the periodic table, as can be seen from fig. 0.2.

The Nature of Science and Priority Disputes

Whereas theories and concepts that appear in textbooks are presented as being fully formed, real science is in a constant state of flux. When science is reported in the press, one seldom hears of the errors that led up to a discovery. In fact, actual science is full of mistakes and wrong turns. We don't ever reach the "truth." The best we can hope for is an approach to the truth, perhaps in an incremental fashion, meaning that current science is necessarily incorrect.

To better understand science is to face up to the historical twists and turns and the mistakes. Moreover, the practice of science often involves struggles between individuals or teams of scientists trying to establish their priority, not because scientists are egotists, although some are, but because scientific society rewards the winners and those who can boldly assert their claims. In the search to discover elements, priority disputes have frequently occurred and in some cases continue to occur to this day. One of the most bitter priority issues involved the discovery of element 72, which was eventually named hafnium as we will see.

Heated arguments and protracted debates, often with nationalistic undertones, are part of science whether we may like it or not. In fact, scientific knowledge as a whole might be said to benefit from the fierce scrutiny to which new claims are subjected, even if the individuals involved in the process may suffer in the process. Scientific knowledge, as it develops, is not in the slightest bit interested in the feelings of individual scientists. What matters is progress in overall human knowledge and not whether the rewards go to one or other person or nation. Having said that, scientists are humans and scientific knowledge is influenced by various more emotional aspects.

Questions of priority are rampant in science. Consider the case of Darwin versus Wallace, Newton versus Leibniz, Montagnier versus Gallo, or Venter versus Collins. In chemistry, there have been cases like Lothar Meyer versus Mendeleev, Ingold versus Robinson, and H. C. Brown versus Winstein. The present book will explore the priority disputes over the discovery of some elements since the start of the twentieth century up to the present time. In some senses the issue should be more clear-cut in the case of elements. One would think that a scientist either did or did not discover a particular element when they claimed they did. Unfortunately, as we will see, this is seldom the case. In some claimed discoveries, it is possible that an individual might have isolated a particular element but might not have characterized it sufficiently well.

An interesting aspect of priority disputes is that people not directly involved in the research often take up the cause of a particular scientist and pursue it to the hilt. This has been the case in many of the elements discussed in this book such as hafnium, technetium, rhenium, and promethium. In the case of hafnium it was the scientific and popular press who seem to have made the most vociferous statements on behalf of one or other of the parties who claimed to have discovered the element. To this day there are several chemists and physicists claiming that some of the elements mentioned above were discovered in the early twentieth century and that they failed to gain credit for one reason or another.

Of course, science is also replete with simultaneous discoveries such as happened in the case of the periodic table itself, the discovery of which is usually attributed to Mendeleev alone.[15] We have to be prepared for the possibility that many elements were also simultaneously discovered.

As Joel Levy points out in his charming book about scientific feuds,[16]

> The history of science is boring; the traditional version, that is, with its stately progression of breakthroughs and discoveries, inspirational geniuses and long march out of the darkness of ignorance into the light of knowledge. This is the story as it is often presented in museums, textbooks and classrooms; but it is an invention...

Levy's book, as well as mine, aims to tell a more interesting, more accurate, and hopefully more satisfying account of what actually happened.

A Sociologist of Science on Priority Disputes

Surprisingly little seems to have been written with the aim of analyzing the nature of priority disputes in modern science. One important exception is the work of the leading sociologist of science of the classical tradition, namely Robert Merton.[17] In an article written in 1957, he says:

> We begin by noting the great frequency with which the history of science is punctuated by disputes, often by sordid disputes, over priority of discovery. During the last three centuries in which modern science developed, numerous scientists, both great and small, have engaged in such acrimonious controversy.

He continues by mentioning that far from being a rare exception in science, priority disputes have long been "frequent, harsh, and ugly" and that they have practically become an integral part of the social relations between scientists.

It would seem a simple matter for scientists to concede that simultaneous discoveries can often occur and that the question of priority is consequently beside the point. On some occasions, this is just what has happened, as in the cases of Darwin and Wallace, who tried to outdo one another in giving credit to the other for their discoveries. Fifty years after the event, Wallace was still insisting upon the contrast between his own hurried work, written within a week after the great idea came to him, and Darwin's work, based on twenty years of collecting evidence.

While Wallace admitted that he had been a "young man in a hurry," Darwin had been a "painstaking and patient student seeking ever the full demonstration of the truth he had discovered, rather than to achieve immediate personal fame." Merton recounts how in some cases, self-denial has gone even further. For example, Euler withheld his long "sought solution to the calculus of variations, until the twenty-three-year-old Lagrange, who had developed a new method needed to reach the solution, could put it into print," so as not to deprive Lagrange. Nevertheless, Merton writes, the recurrent struggles for priority, with all their intensity, far overshadow these rare cases of noblesse oblige.

Merton turns to several possible explanations for the almost ubiquitous current state of common priority disputes. First of all, priority disputes may be the mere expressions of our essentially competitive human nature. If egotism is natural to the human species, scientists will have their due share of egotism and will sometimes express it through exaggerated claims to priority.

A second explanation considered is that like other professions, the occupation of science attracts some, and perhaps even many, ego-centered people who are hungry for fame. Merton quickly dismisses this possibility because he doubts that quarrelsome or contentious personalities are especially attracted to science and recruited into it.

Significantly, Merton recognizes that priority disputes often involve men of ordinarily modest disposition who act in self-assertive ways only when they come to defend their rights to intellectual property. He also points out that often, the principals themselves, the discoverers or inventors, take no part in arguing their claims to priority or perhaps withdraw from the controversy. Instead, it is their followers who commonly take up the cause of some particular scientist and begin to regard the attribution of priority as a moral issue that must be fought-out.[18]

But why should these supporters have any interest in defending the claims of a particular scientist? Merton believes that by identifying themselves with the scientist or with the nation of which they are a part, they can somehow partake of the glory that is to be obtained if the fight is won. In the present book we will see many cases of priority disputes involving the discovery of several elements and in the case of element 72, some disputes conducted by supporters of the rival claimants.

While Merton believes that neither human nature nor the role of supporters explains priority conflicts, he considers the question of institutional norms to be far more significant.

> As I shall suggest, it is these norms that exert pressure upon scientists to assert their claims, and this goes far toward explaining the seeming paradox that even those meek and unaggressive men, ordinarily slow to press their own claims in other spheres of life, will often do so in their scientific work.[19]

Next Merton turns to make what I believe is his most astute point, when he begins to talk of scientific knowledge as a form of property. In a commercial setting the protagonists in a dispute can often resolve their differences because there is money to be made from the property in question. But in academic life a discovery leads to intellectual property, which is seldom commercially exploitable—or at least this was the case for Merton, who was writing in 1957. As a result, the *only* thing that the scientist can benefit from his or her "property" is the fame from having discovered the knowledge. Small wonder then that scientists will fight so ferociously to retain the only benefits that might come from their hard-won intellectual property. As Merton puts it,

> Once he has made his contribution, the scientist no longer has exclusive rights of access to it. It becomes part of the public domain of science. Nor has he the right of regulating its use by others by withholding it unless it is acknowledged as his. In short, property rights in science become whittled down to just this one: the recognition by others of the scientist's distinctive part in having brought the result into being.

Merton, of course, recognizes the element of nationalism in priority disputes when writing,

> From at least the seventeenth century, Britons, Frenchmen, Germans, Dutchmen, and Italians have urged their country's claims to priority; a little later, Americans entered the lists to make it clear that they had primacy.

In the episodes discussed in the present book, nationalism seems to have played an unusually large role. The prime example was the discovery of hafnium, which split the scientific community along the militaristic lines of World War I, which had only recently ended. British and French journals, and newspapers especially, took the side of Urbain, who claimed to have discovered element 72 in the form of celtium. Meanwhile, the opposition, which consisted of Coster and Hevesy working in Bohr's laboratory in Copenhagen, Denmark, were associated with the wartime enemy of Germany and Austria. Another more recent case of such politically motivated nationalism occurred in the case of the synthesis of element 105 in the 1960s when US and Soviet scientists bickered over which of them had been the first to create the element that was to eventually become dubnium, after Dubna in Russia.[20] As an interesting contrast to this period of Cold War detente, the American scientists gave element 101 the name of mendelevium, thus honoring a scientist from the "opposing side."

One rather extreme situation that results from the desire for priority on the part of the scientist is the occasional surrender to the urge to falsify scientific findings. Interestingly, we have witnessed a case of this form of activity as in the synthesis of element 118, over which a senior Berkeley scientist was dismissed from his position for allegedly falsifying data.

A less outrageous form of behavior, which is rather more common, is for scientists to fail to cite their competitors. This was true of Mendeleev, the leading discoverer of the periodic table, as many authors have pointed out. Whereas Mendeleev was content to cite the articles of early researchers such as Döbereiner and Pettenkofer, he seems to have been more reluctant to acknowledge the work of immediate competitors such as Lothar Meyer and Newlands, whose work he criticized rather severely. In fact, Mendeleev conducted a rather acrimonious priority dispute with Meyer.

Work on Discovery and Priority since Merton

Much work has been conducted within the sociology of science and "science studies" following the work of Merton and it is necessary to turn briefly to this research.[21] In doing so I will concentrate on an influential article by Alan Gross, published in 1998.[22] Gross writes,

> I will show that the normative requirements for calling an event a scientific discovery are such that priority and the conflicts it generates are not merely *in* science, but *of* science.

Stated in other words, Gross is trying to highlight the importance of priority debates in science, which scientists often try to deny or at least play down as being something that intrudes into the scientific landscape rather than being an integral part of the nature of science.

He also claims that discovery is not a historical event but rather a retrospective social judgment.

This again is an important point worth bearing in mind as we examine the discovery of the last seven elements that fell into place in the old periodic table. Gross mentions that science historian Thomas Kuhn had already pointed out that scientific discovery cannot be regarded as a historical event such as a war, a revolution, or the crash of the stock exchange. However, Gross believes that Kuhn does not go far enough in recognizing the essentially social role played in discoveries.

What Exactly Is a Discovery of an Element?

Since this book is largely about the discovery of seven elements, it is necessary to consider what actually constitutes the discovery of any element as opposed to a scientific discovery in general.[23] Many popular books on the subject try to settle the question in favor of one particular individual or team of researchers. In fact, many cases cannot be easily settled and any attempt to narrow the decision down to one individual deprives us of the richness of the story of any particular element.[24] It also masks the fact that the discovery of an element, and indeed any scientific discovery, is frequently a long, drawn-out process involving many individuals over many stages. It is often not easy to pinpoint just one crucial discovery.

In the case of elements, some specific problems emerge that may not be common to other types of scientific discovery. For example, is it necessary for an individual to have recognized that he or she had isolated an element or is it enough that they had isolated what they believed to be merely a new substance that eventually was recognized as an element?

Moreover, is it necessary for the element to have been isolated or is it sufficient for a scientist to have realized that it was present in a certain mineral? And even if an individual may have succeeded in isolating a new element, while fully realizing that it was indeed an element, does it matter how much was isolated? Is it essential to have a visible amount of the element to claim its legitimate discovery? If this were the case it would immediately rule out most of the transuranium elements since many of these have only been produced in miniscule amounts. In some cases it may have been as little as two or three atoms! In fact, this very problem crops up, at least in principle, in several of the seven elements that fall within the first ninety-two elements. One might call these elements the infra-uranium elements to distinguish them from the transuranium variety.

There are also other issues that might complicate any attempt to identify any one particular discoverer or team of discoverers. How must the discovery of the putative element have been announced to other scientists and the world at large? Perhaps we should favor announcements that give enough detail to allow other researchers to repeat the procedure. But should we deprive an author from being regarded as the official discoverer if he omitted to go into such details as to allow others to repeat the procedure? After all, the author may have been prevented from doing so by a journal editor who is keen on saving page space.

More importantly perhaps, should we insist that the announcement be made in the scientific literature rather than a magazine, a newspaper, or a TV or radio show?[25] What if the discovery was not officially announced anywhere but was later found described in a scientist's notebook long after the discovery had been attributed to a separate event that had taken place later? Should the roster of official discovery of the elements be revised because of this notebook? But, as Alan Gross points out,

> To publish is to take that risk, to oblige oneself, and one's intellectual allies and heirs, to defend one's claims in public.

Must a newly discovered element be announced in a reputable scientific journal? If so then the discovery of palladium, for example, could not be rightly attributed to Wollaston, who famously chose to announce the discovery in a newspaper advertisement.

A separate question relates to what level of purity we should demand before accepting a particular piece of research as a genuine discovery. Kuhn put the matter rather concisely when he wrote,

> any attempt to date the discovery must inevitably be arbitrary because discovering a new sort of phenomenon is necessarily a complex event, one which involves recognizing both *that* something is and *what* it is. [26]

In some cases an element is isolated but may remain unknown until another person has isolated and published their own results. Who should be considered as the real discoverer or should we even try to settle such delicate issues? The discovery of oxygen represents a notoriously complicated case that has been the cause of much debate over the centuries and that has even been the subject of a recent stage play.[27] Oxygen was isolated by the Swedish chemist Scheele, who was probably not the first person to do so. It was then independently rediscovered by Priestly in England sometime later. Priestly visited Lavoisier in Paris and told him about this work. Lavoisier then rediscovered oxygen but went much further than his contemporaries in making this element the centerpiece for a new theory of combustion that overthrew the notion that burning resulted in the evolution of a substance called phlogiston. Should Lavoisier get the credit for the discovery of oxygen because he put this discovery to much greater use than had Priestly or Scheele?[28]

Another interesting case concerns the element platinum but for a somewhat different reason. Its discovery is frequently attributed to one or more European scientists who either visited America or received a sample from that continent. And yet it is clear that the indigenous people of the Americas had discovered platinum prior to these events.

Many elements, including platinum perhaps, might best be considered as having been anonymously discovered. They include the seven metals of antiquity, along with carbon and sulfur. Similarly, some elements discovered before the year 1500 cannot be assigned to any particular individual. These elements include arsenic, antimony, bismuth, and zinc, although none of these were regarded as elements at the time.[29] It appears that the first element whose discovery can be assigned to a specific person is phosphorus. This discovery is almost universally attributed to Hennig Brand, even though he did not even publish his finding.

A Philosophical Interlude on Elements

In the philosophy of science, Karl Popper's views on refutation have become rather influential. It has long been known that scientific theories and laws cannot be proved. Popper points out that the logical asymmetry between proof and disproof, which he calls refutation, can be put to good advantage. In the case of proof this can never be achieved by the accumulation of evidence since it is always possible for future discoveries to contradict a theory or law, however well established it might seem to be on the basis of past evidence. On the other hand, a single refutation is, in principle at least, capable of establishing that "not all swans are white" or of refuting a theory or law. Should this criterion be extended to the discovery of elements? This is not a question that has been discussed before as far as I am aware.

On the other hand, the application of Popper's idea to the discovery of elements might be going too far. After all, there is supposed to be a clear-cut criterion for the identification of elements. It lies in the concept of atomic number, which is taken to be a unique indicator of any particular element. However, the implementation of this criterion is a more complicated matter. How do we actually know if we have observed a substance with a new atomic number? The answer is supposed to reside in the use of Moseley's classic X-ray method but this approach is fraught with problems. For example, the observation of faint X-ray lines has caused the false identification of certain elements such as element 43—one of the seven elements that are the focus of this book. In 1925 the Noddacks and Berg believed that they had isolated this element and tried to confirm their discovery by recording what they claimed to be the X-ray spectrum of the new substance.

In fact, one of the main criticisms of Popper's approach to philosophy of science concerns just such issues. According to the Duhem-Quine thesis, any attempt to test a theory or law always remains ambiguous because the process of testing complicates the issue. One cannot directly test the theory or law. The testing must be carried out using some auxiliary assumptions, which serve to spell out the theory or law in any particular context. Consider, for example, the outcome of dipping half the length of a pencil in a glass of water. The result is that the pencil will appear to be bent due to the effect of refraction.

Suppose that in another experiment a stick does not show such familiar bending. Are we to conclude that the law of refraction has been refuted? Not necessarily, since the failure to observe the expected behavior may lie in the initial conditions rather than with a failure of a scientific law. It may be that, on this occasion, the stick has been dipped into some other liquid for which the bending is minimal or nonexistent.

Thomas Kuhn on Discovery

The historian of science Thomas Kuhn is rightly famous for having published one of the influential books of the second half of the twentieth century, entitled *The Structure of Scientific Revolutions.* Just before doing so, in the same year of 1962, Kuhn published a journal article in the magazine *Science,* which is not as well known but very relevant to the contents of the present book given that he examined the nature of scientific discovery.[30]

One of the main proposals that he makes in the article is that there are essentially two kinds of scientific discoveries. First, there are the unexpected discoveries, those that could not have been predicted from theory, among which Kuhn includes the discovery of oxygen, X-rays, and the electron.

His second category consists of those discoveries that were in fact predicted from existing theories, including such examples as the neutrino, radio waves, and missing elements in the periodic system. Kuhn then makes a further claim when he writes,

> As a result there have been *few* priority debates over discoveries of this second sort and only a paucity of data can prevent the historian from ascribing them to a particular time and place (Kuhn, 1962).

The best way I can think of reacting to this statement is to borrow the title of a popular book on a legendary Caltech physicist and say, Surely you're joking, Mr. Kuhn!

As I hope to show in this book, the discovery of the missing seven elements is replete with priority debates. Moreover, it is not because of any paucity of data that historians have frequently been unable to assign the discovery of these elements to any particular place. In fact, as I will also document, several scientists have very recently attempted to rehabilitate some older claims for the discovery of as many as three or four of these elements, including elements 43, 61, and 75.[31]

It's Also a Question of Abundance

One of the main factors responsible for the fact that these seven elements were the last to be discovered is the question of their very low natural abundance, with the possible exception of rhenium. But why the particular atomic numbers in question, namely 43, 61, 72, 75, 87, 85, and 91?

With the exception of Hf, whose atomic number is 72, all the others have atomic numbers with odd rather than even values. This can be understood, roughly speaking, as being due to the fact that the

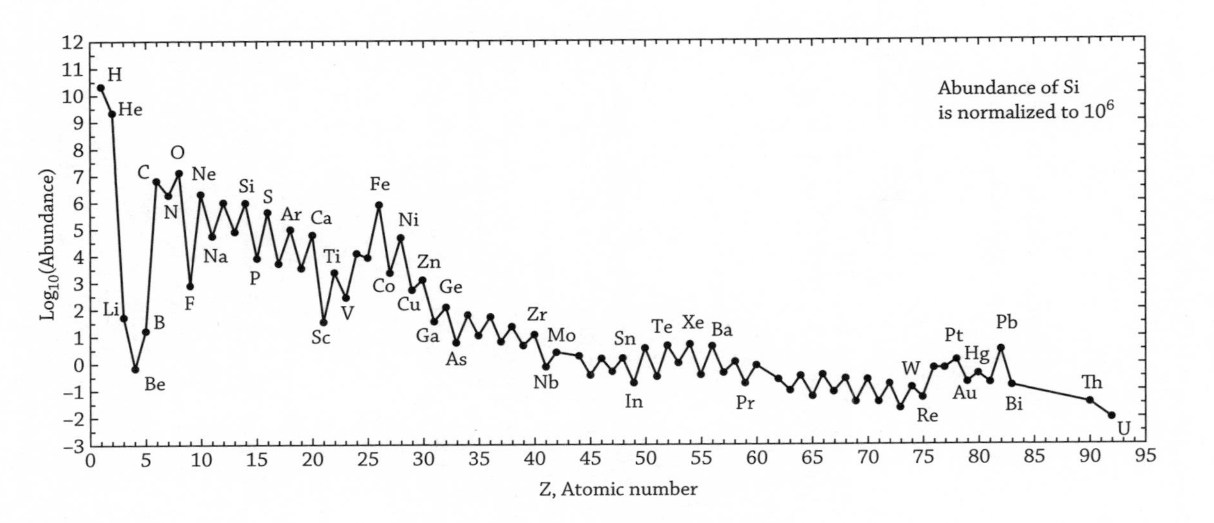

FIGURE 0.3 Relative terrestrial abundance of the elements. Reproduced from P. A. Cox, *The Elements*, Oxford University Press, Oxford, 1989, figure from p. 17 (with permission).

protons in the nucleus of the atoms of these elements can achieve greater stability if they occur in pairs in nuclear energy levels that are analogous to electronic energy levels into which electrons enter up to two at a time.

A plot of the relative abundance of elements clearly shows a sawlike appearance due to this alternating stable (even Z) and less stable (odd Z) behavior (fig. 0.3).

This feature, known as Harkin's rule, goes some way toward explaining why the elements 43, 61, 85, and 87 in particular have such extremely low abundances, namely Tc (almost completely absent), Pm, 10^{-19} percent, Fr, 1.3×10^{-21} percent, At, 3×10^{-24} percent of the Earth's crust.[32]

Questions of Priority

A very important aspect of the discovery stories of many elements has been the question of priority, which has frequently been bitterly disputed. In fact, one would be hard put in trying to find any elements whose discovery has not been a matter of dispute, frequently by researchers from different countries. Exploring this issue gives us an interesting and neglected glimpse at an aspect of the sociology of science. Many authors have asked themselves why discoveries in general and of elements in particular have led to such conflicts. The answers offered are revealing of the often frail humanity of the scientists involved or perhaps of the scientific method as a whole and the pressures it places on scientists.

It so happens that all of the seven elements that form the core of this book have been the subject of priority debates and most of them have been dominated by nationalistic drives. Sometimes this comes from the scientists themselves while in other cases it is supporters of these scientists or perhaps even the press in the countries involved that fuel the nationalism. A perfect example will be seen in the case of element 72, or hafnium, which resulted in one of the most bitter and, at times, most comical examples of scientific nationalism on record.

Before launching into the tale of seven elements, however, we need to explore the discovery of the framework that situates and connects all of the elements, namely the periodic table. Even before doing this we will examine the steps that had to be taken before the periodic table could begin to take shape at the hands of at least six scientists in different parts of the world. The leading discoverer of the periodic table was the Russian chemist Mendeleev, who was in fact the last of the six discoverers, although it was Mendeleev who made the most significant contribution to this icon of science. We will also consider the influence that discoveries made in physics had on explaining the periodic table, including Moseley's work, which established the importance of atomic number.

The reader should note that the following two chapters are condensed versions of most of the chapters of my 2007 book on the periodic table. You may, therefore, wish to skip to chapter 3, which begins relating the discovery of the seven elements.[33]

Chapter 1

From Dalton to the Discovery of the Periodic System

Our story begins, somewhat arbitrarily, in the English city of Manchester around the turn of the nineteenth century. There, a child prodigy by the name of John Dalton, at the tender age of fifteen is teaching in a school with his older brother. Within a few years, John Dalton's interests have developed to encompass meteorology, physics, and chemistry. Among the questions that puzzle him is why the various component gases in the air such as oxygen, nitrogen, and carbon dioxide do not separate from each other. Why does the mixture of gases in the air remain as a homogeneous mixture?

As a result of pursuing this question, Dalton develops what is to become modern atomic theory. The ultimate constituents of all substances, he supposes, are hard microscopic spheres or atoms that were first discussed by the ancient Greek philosophers and taken up again by modern scientists like Newton, Gassendi, and Boscovich. But Dalton goes a good deal further than all of these thinkers in establishing one all-important quantitative characteristic for each kind of atom, namely its weight. This he does by considering quantitative data on chemical experiments. For example, he finds that the ratio for the weight in which hydrogen and oxygen combine together is one to eight. Dalton assumes that water consists of one atom of each of these two elements. He takes a hydrogen atom to have a weight of 1 unit and therefore reasons that oxygen must have a weight of 8 units. Similarly, he deduces the weights for a number of other atoms and even molecules as we now call them (fig. 1.1).

Element	Weight
Hydrogen	1
Azot	4.2
Carbon (charcoal)	4.3
Ammonia	5.2
Oxygen	5.5
Water	6.5
Phosphorus	7.2
Nitrous gas	9.3
Ether	9.6
Nitrous oxide	13.7
Sulphur	14.4
Nitric acid	15.2

FIGURE 1.1 Part of an early table of atomic and molecular weights published by Dalton. J. Dalton, *Memoirs of the Literary and Philosophical Society of Manchester*, 2, 1, 207 (1805).

For the first time the elements acquire a quantitative property, by means of which they may be compared. This feature will eventually lead to an accurate classification of all the elements in the form of the periodic system, but this is yet to come. Before that can happen the notion of atoms provokes tremendous debates and disagreements among the experts of Dalton's day. Some, like Dalton's compatriot, Thomas Thomson, take up the notion of atoms very enthusiastically. Others, like Berzelius in Sweden and Dumas in France, are more reluctant. Some argue that since there is no direct evidence for these microscopic particles, atoms should be regarded only as a useful instrument and not as literally describing reality.[1]

In addition, there is disagreement over the concepts of atoms and molecules. Some use the terms interchangeably. Two notable exceptions are the Italian Amedeo Avogadro, and the Frenchman Ampère, but nobody takes them very seriously. For example, Avogadro believes that the simplest part of many gases is a molecule consisting of two atoms that are bonded together. Dalton, among others, refuses to accept this notion, arguing that two atoms of the same element would repel each other and so cannot possibly combine together. It turns out that Avogadro is correct on this point but it takes until the year 1860 before a full clarification of terms like atom and molecule are hammered out. At about the same time it is realized that a molecule of water contains two atoms of hydrogen and one atom of oxygen to give H_2O, again contrary to Dalton's formula of HO. As a result of this change in perspective concerning molecules and atoms, which was brought about by another Italian, Cannizzaro, many of Dalton's atomic weights become altered. Oxygen atoms are now reckoned to have a weight of 16 and not 8 as Dalton had originally asserted.

A new set of atomic weights is distributed by Cannizzaro in the form of a pamphlet at the first international chemical congress held in Karlsruhe, Germany. Soon afterward, it is published in the scientific literature for all to see. Within a period of approximately ten years, as many as six or more scientists in various countries independently arrive at periodic systems of the elements by using Cannizzaro's revised atomic weights as a means of ordering the elements.

But a periodic system is not just a linear sequence of elements arranged in order of increasing atomic weights. What it also provides is a display of the approximate repetition in the properties of the elements that seems to occur after certain regular intervals or periods. These chemical similarities had been known even before Dalton had devised his first set of atomic weights. They include elements such as lithium, sodium, and potassium, all soft gray metals that react readily with water to produce hydrogen. Another example consists of chlorine, bromine, and iodine, which show marked similarities in forming acids when combined with hydrogen and forming colorless crystalline salts when combined with a metal such as sodium.

De Chancourtois

The discovery of the first periodic system is made by the unknown, and for some considerable time unrecognized, geologist, Alexandre-Emile Béguyer de Chancourtois. What de Chancourtois does is to arrange the elements on a helical spiral drawn on the surface of a three-dimensional cylinder. The chemically similar elements like lithium, sodium, and potassium are found on vertical lines that are made to intersect the helical arrangement of elements as shown in fig. 1.2.

In this way, de Chancourtois appears to take perhaps the single most important step in the discovery of the periodic system. It is he who first recognizes that the properties of the elements are a periodic function of their atomic weights, a full seven years before Mendeleev arrives at the same conclusion.

But de Chancourtois is not generally accorded very much credit, partly because his publication does not appear in a chemistry journal and partly because he fails to develop his insight any further over subsequent years. Only about thirty years after his paper appears does de Chancourtois's claim to priority come to light because of some supporters in England and France.[2]

De Chancourtois arranges the elements according to what he calls increasing "numbers" along a spiral. These numbers are written along a vertical line that serves to produce a vertical cylinder. The circular base of the cylinder is divided into sixteen equal parts. The helix is traced at an angle of 45 degrees to its vertical axis and its screw thread is divided, at each of its turns, into sixteen portions. Thus, the seventeenth point along the thread lies directly above the first, the eighteenth above the second, and so on. As a result of this representation, elements whose characteristic numbers differ by sixteen units become aligned in vertical columns. For example, sodium, with a weight of 23, appears one complete turn above lithium, whose value is taken as 7. The next column contains the elements magnesium, calcium, iron, strontium, uranium, and barium.

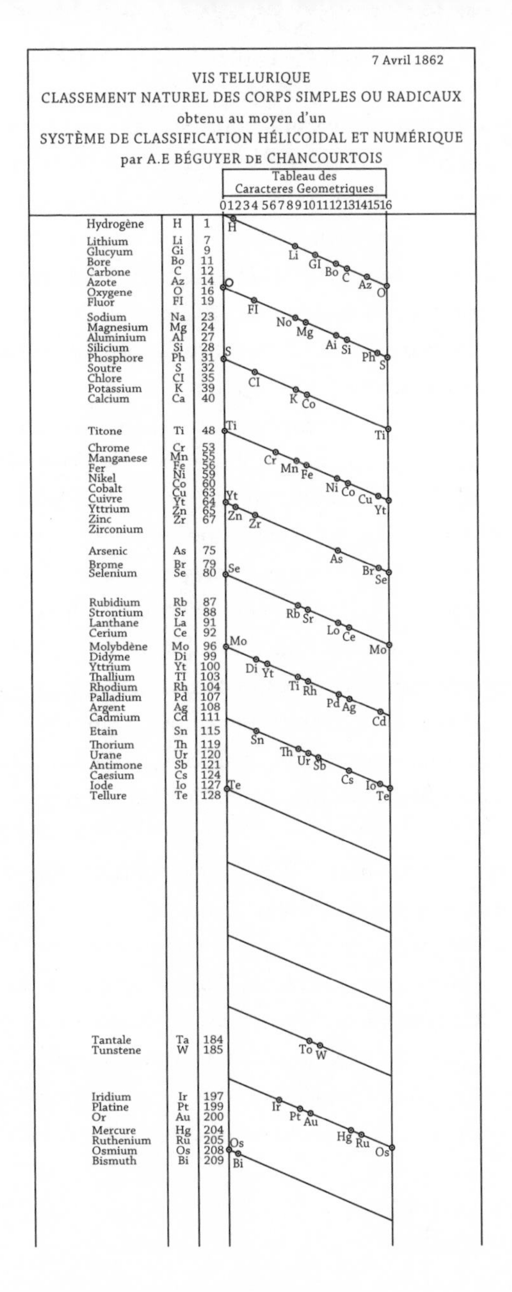

FIGURE 1.2 De Chancourtois helical periodic system. A. E. Béguyer de Chancourtois, Vis Tellurique: Classement naturel des corps simples ou radicaux, obtenu au moyen d'un système de classification hélicoïdal et numérique, *Comptes Rendus de L'Académie des Sciences*, 54, 757–761, 840–843, 967–971, 1862.

The first full turn of the spiral ends with the element oxygen and the second full turn is completed at sulfur. Periodic relationships, or chemical groupings, can therefore be seen in de Chancourtois's system, although only approximately, by moving vertically downward along the surface of the cylinder. The eighth such turn, which is also the halfway point down the cylinder, occurs at tellurium. This rather arbitrary feature provides de Chancourtois with the name of "vis tellurique," or telluric screw, for his system.[3]

De Chancourtois's system fails to create any impression on chemists because the original published article does not include a diagram, mainly due to incompetence on the part of his publisher. Another problem is that the system does not convey chemical similarities very convincingly because of the unusual spiral representation. While some of the intended chemical groupings, like the alkali metals, the alkaline earths, and the halogens fall into vertical columns, many others do not, thus making it a less successful system than it might have been. Yet another drawback to the system is the inclusion of radicals such as NH_4^+ and CH_3, compounds like cyanogen, and even some alloys, none of which belong in a system of the elements.

Frustrated that the journal *Comptes Rendus* has failed to include a diagram, de Chancourtois privately republishes his system in 1863. But, because it is published privately, this further article receives even less attention from other scientists.[4] Nevertheless, there is no denying that de Chancourtois is the first to show that the properties of the elements are a periodic function of their atomic weights, or to cite his own words: "Les proprietées des corps sont les proprietées des nombres." [The properties of bodies are the properties of numbers].[5]

John Newlands

John Newlands is born in 1837 in Southwark, a suburb of London, which by coincidence is also the birthplace of William Odling, another pioneer of the periodic system. After studying at the Royal College of Chemistry in London, Newlands becomes the assistant to the chief chemist of the Royal Agricultural Society of Great Britain. In 1860 he serves briefly as an army volunteer with Garibaldi, who is fighting the revolutionary war in Italy. The reason for Newlands's willingness to fight in this war is apparently due to his mother's Italian descent. It also means that Newlands is unable to attend the Karlsruhe conference of the same year, although since he is not a major chemist at the time he would probably not be invited. After returning to London, Newlands begins working as a sugar chemist, while also supplementing his income with private chemistry lessons.

Newlands's first attempt at classification concerns a system for organic compounds that he publishes in 1862 together with proposals for a new system of nomenclature. In the following year, he publishes the first of many classification systems for the elements. In 1863, when Newlands puts forth his first system, he does so without the benefit of the atomic weight values that had been issued following the Karlsruhe conference of 1860. Instead, he uses the atomic weight values favored by Gerhardt, who had begun to revise atomic weights even before the Karlsruhe conference. Newlands is able to produce a table consisting of eleven groups of elements with analogous properties whose weights differ by a factor of 8 or some multiple of 8.

In his 1863 article Newlands describes a relationship among atomic weights of the alkali metals and uses it to predict the existence of a new element of weight 163, as well as a new element that would occur between iridium and rhodium. Unfortunately for Newlands, neither of these elements is ever discovered.[6]

In 1864, Newlands publishes a second article on the classification of the elements (fig. 1.3). This time, he draws on the more correct, post-Karlsruhe, atomic weights. Newlands now finds a difference of 16, or very close to this value, instead of 8, between the weights of six sets of first and second members among groups of similar elements.

This finding seems unexpectedly accurate given that he is working with atomic weights and not atomic numbers.[7]

Less than a month after his first system of 1864 appears, Newlands releases yet another system, (fig. 1.4) but with fewer elements (24, plus a space for a new element) and makes no mention of atomic weights. The article is nevertheless of considerable merit since Newlands assigns an ordinal number to each of the elements, thus, at least superficially, appearing to anticipate the modern notion of atomic number. Abandoning the arithmetic progressions in atomic weights that had bedeviled earlier investigators, Newlands simply arranges the elements in order of increasing atomic weight without concern for the values of those weights.

The most important thing Newlands does in his third publication on the classification of the elements is to present a periodic system—that is, he reveals a pattern of repetition in the properties of the elements after certain regular intervals. This is the essence of the periodic law, and Newlands, along with de Chancourtois, deserves more credit for having recognized this fact so early.

Member of a group having lowest equivalent		One element immediately above the preceding one		Difference	
				H = 1	O = 16
magnesium	24	calcium	40	16	1
oxygen	16	sulphur	32	16	1
lithium	7	sodium	23	16	1
carbon	12	silicon	28	16	1
fluorine	19	chlorine	35.5	16.5	1.031
nitrogen	14	phosphorus	31	17	1.062

FIGURE 1.3 Newlands's first table of 1864. (The original spelling of "sulphur" has been retained.) Redrawn from J. A. R. Newlands, *Chemical News*, 10, 59–60, 1864. Table on p. 59.

			No.		No.		No.		No.		No.
Group	*a*	N	6	P	13	As	26	Sb	40	Bi	54
"	*b*	O	7	S	14	Se	27	Te	42	Os	50
"	*c*	Fl	8	Cl	15	Br	28	I	41	—	—
"	*d*	Na	9	K	16	Rb	29	Cs	43	Tl	52
"	*e*	Mg	10	Ca	17	Sr	30	Ba	44	Pb	53

FIGURE 1.4 Newlands's table of 1864. from J. A. R. Newlands, Relations Between Equivalents, *Chemical News, 10,* 59–60, 1864. Table on p. 59.

The Law of Octaves

In 1865, Newlands develops yet another system, which is an improvement on that of the previous year because he now includes sixty-five elements, in increasing order of atomic weight, while once again using ordinal numbers rather than actual values of atomic weight. This system is built upon his famous "law of octaves" whereby the elements show a repetition in their chemical properties after intervals of eight elements.[8] Newlands goes so far as to draw an analogy between a period of elements and musical octaves, in which the tones display a repetition involving an interval of eight notes (counting from one note of C, for example, to the next note C inclusive). To cite the words of Newlands himself:

> If the elements are arranged in the order of their equivalents with a few slight transpositions, as in the accompanying table, it will be observed that elements belonging to the same group usually appear on the same horizontal line. It will also be seen that the numbers of analogous elements differ either by 7 or by some multiple of seven; in other words, members of the same group stand to each other in the same relation as the extremities of one or more octaves in music... The eighth element starting from a given one is a kind of repetition of the first. This particular relationship I propose to term the *Law of Octaves*.[9] [Original italics]

This statement marks a rather important step in the evolution of the periodic system since it represents the first clear announcement of a new law of nature relating to the repetition of the properties of the elements after certain intervals. Whether Newlands seriously intends to suggest a connection between chemistry and music is not clear. In any case, his fanciful analogy is not the sole reason why the chemists in attendance are quick to dismiss Newlands's scheme. Their hostility is better attributed to the British tendency of the time to be suspicious of theoretical ideas in general. The best-known response to Newlands is the much-quoted one of George Carey Foster, who suggests that Newlands might obtain a superior classification if he were to merely order the elements alphabetically!

The chemists gathered at the London Chemical Society meeting decide not to permit publication of Newlands's article in the society's proceedings. Undaunted, Newlands publishes several further articles in the highly respected journal *Chemical News*, including the contents of his presentation to the London Chemical Society.

In an article published in 1866, Newlands tries to answer the criticisms that were leveled at him in the course of his presentation to the Chemical Society. The table accompanying the article represents the first time that Newlands arranges chemical groups in vertical columns. Once again the ordering of the elements follows a numerical sequence, with the exception of three reversals (Ce & La with Zr; U with Sn; and Te with I). In the following quote, Newlands responds to a criticism that he had not left any gaps and that this would be a problem when future elements were discovered.

> The fact that such a simple relation [the law of octaves] exists now, affords a strong presumptive proof that it will always continue to exist, even should hundreds of new elements be discovered. For, although the difference in the number of analogous elements might, in that case, be altered from 7, to a multiple of 7, of 8, 9, 10, 20, or any conceivable figure, the existence of a simple relation among the elements would be none the less evident.[10]

Newlands is, of course, correct.[11] He is vindicated by the subsequent discovery of the noble gases, which, instead of disrupting the repeating pattern, simply increases the repeat distance between successive periods to 8 rather than 7.[12] There can be no doubt that Newlands ranks among the true pioneers of the modern periodic system, in particular for being the first to recognize explicitly the existence of the Periodic Law, which in many ways is the real crux of chemical periodicity.

William Odling, the Person

Unlike many of the discoverers of the periodic system, who are marginal figures in the history of chemistry, William Odling is a distinguished chemist and scientist who holds some important positions in the course of his career. Most notably, he succeeded Michael Faraday as director of the Royal Institution in London. Odling holds the distinction of having attended the Karlsruhe conference, where he gives a lecture on the need to adopt a unified system of atomic weights. Unlike Newlands, whose first attempts at a periodic system are carried out in ignorance of Cannizzaro's recommended values of atomic weights, Odling is able to use these values from the beginning of his attempts at producing a table of the elements. Following the Karlsruhe meeting, Odling becomes the leading champion of the views of Cannizzaro and Avogadro in England. Odling above all others, therefore, recognizes the significance of the new atomic weight values for the classification of the elements.

His main article on the periodic system appears in 1864, while he is a reader in chemistry at St. Bartholomew's Hospital in London. Whereas Newlands's system of the same year includes only twenty-four of the sixty known elements, Odling succeeds in including fifty-seven of them (fig. 1.5). Furthermore, Odling's paper precedes Newlands's announcement of chemical periodicity to the London Chemical Society. Nevertheless, it appears that the two chemists are working completely independently of each other.

			Ro 104	Pt 197
			Ru 104	Ir 197
			Pd 106.5	Os 199
H 1	„	„	Ag 108	Au 196.5
„	„	Zn 65	Cd 112	Hg 200
L 7	„	„	„	T1 203
G 9	„	„	„	Pb 207
B 11	A1 27.5	„	U 120	„
C 12	Si 28	„	Sn 118	„
N 14	P 31	As 75	Sb 122	Bi 210
O 16	S 32	Se 79.5	Te 129	„
F 19	Cl 35.5	Br 80	I 127	„
Na 23	K 39	Rb 85	Cs 133	„
Mg 24	Ca 40	Sr 87.5	Ba 137	„
	Ti 50	Zr 89.5	Ta 138	Th 231.5
	„	Ce 92	„	
	Cr 52.5	Mo 96	V 137	
	Mn 55		W 184	
	Fe 56			
	Co 59			
	Ni 59			
	Cu 63.5			

FIGURE 1.5 William Odling's periodic system. W. Odling, On the Proportional Numbers of the Elements, *Quarterly Journal of Science, 1*, 642–648, 1864, p. 643.

Odling begins his article with:

Upon arranging the atomic weights or proportional numbers of the sixty or so recognized elements in order of their several magnitudes, we observe a marked continuity in the resulting arithmetical series...[13]

Then he makes an observation that amounts to an independent discovery of the periodic system:

> With what ease this purely arithmetical seriation may be made to accord with a horizontal arrangement of the elements according to their usually received groupings, is shown in the following table, in the first three columns of which the numerical sequence is perfect, while in the other two the irregularities are but few and trivial:[14]

The fact that Odling recognizes the periodicity in chemical properties is evident from the horizontal groupings in this table. He notes that there are a considerable number of pairs of chemically analogous elements, indeed half of all the known elements, whose difference in atomic weights lies between the values of 84.5 and 97. Some of these pairs are shown in fig. 1.6.

Odling also notices that about half of the above cases include the first and third members of previously known triads. He suggests that a middle member might be found for the other half, stating that, "the discovery of intermediate elements in the case of some or all of the other pairs is not altogether improbable." Here then is an example of a prediction made on the basis of a periodic system, although a rather tentative one that is not further developed to any extent.

Odling makes the further claim that the chemical similarities between elements separated by differences of about 48 in atomic weight, such as cadmium and zinc, are greater than those between pairs of elements, such as zinc and magnesium, which are separated

I	–	Cl	or	127	–	35.5	=	91.5
Au	–	Ag		296.5	–	108	=	88.5
Ag	–	Na		108	–	23	=	85
Cs	–	K		133	–	39	=	97

FIGURE 1.6 Odling's first table of differences. W. Odling, *Quarterly Journal of Science*, 1, 642–648, 1864.

by other intervals, such as 41 in this case. Thus it would appear that he recognizes the need to separate certain elements (those that would eventually become known as the transition metals) from the main body of the table. In this way, periodicity can be retained in the properties of the majority of the elements, as is done in the modern medium-long form of the table.[15] If the transition metals are separated out of the short form table, the primary periodic relationship between main-group elements is emphasized and the fact that period lengths vary is accommodated in a natural manner.

From the perspective of the modern periodic table, cadmium and zinc are both transition metals that show a primary kinship. On the other hand, zinc and magnesium belong to the transition metals and main-group elements, respectively, and only show secondary kinships. Odling may thus have anticipated the modern trend to separate zinc and magnesium into different groups and also different blocks of the periodic table.

Gustavus Hinrichs

The case of Gustavus Hinrichs is rather unusual among the discoverers of the periodic system. His scientific interests were so far ranging and so diverse that some commentators regard him as rather eccentric. Hinrichs is born in 1836 in Holstein, a part of Denmark that later becomes a German province. He publishes his first book at the age of twenty, while at the University of Copenhagen. Hinrichs immigrates to the United States in 1861 to escape political persecution. After only a year of teaching high school he is appointed head of modern languages at the University of Iowa and just one year later he becomes professor of natural philosophy, chemistry, and modern languages. He is also credited with founding the first US meteorological station in 1875, and acting as its director for fourteen years.

Hinrichs held a number of academic appointments, first at Iowa State University and later at the University of Missouri, St. Louis, but he seemed to go out of his way to cultivate eccentricity. In addition, he seldom gave references to other authors in his numerous publications. Hinrichs is a prolific author, publishing about three thousand articles in Danish, French, and German as well as in English, in addition to approximately twenty-five books in English and German. These books include the highly eccentric *Atomechanik,* in which Hinrichs gives his views on the classification of the elements. The majority of Hinrichs's articles are published in languages other than English, because he is irritated by American editors who insist on correcting his work and thereby cause delays in publication.

One of the few articles ever written on Hinrichs describes him in the following way:

> It is not necessary to read far into Hinrichs' numerous publications to recognize the marks of an egocentric zeal which defaced many of his contributions with an untrustworthy eccentricity. Only at this late date does it become possible to separate those inspirations which were real—and which swept him off his feet—from background material which he captured in the course of his own learning. Whatever the source, Hinrichs usually dressed it with multilingual ostentation, and to such a point of disguise that he even came to regard Greek philosophy as his own.[16]

But careful consideration of Hinrichs's work shows that there was much useful science, if one is prepared to take time to examine the various strands of his research.

Hinrichs takes a rather Pythagorean approach to science and is captivated by numerical relationships involving very diverse phenomena. By means of a rather ingenious argument, he postulates that atomic spectra can provide information on the dimensions of atoms.[17] Hinrichs's wide range of interests extend to astronomy. Like many authors before him, as far back as Plato, Hinrichs notices numerical regularities regarding the sizes of the planetary orbits. In an article published in 1864, Hinrichs produces the following table (fig. 1.7).

	Distance to the Sun
Mercury	60
Venus	80
Earth	120
Mars	200
Asteroid	360
Jupiter	680
Saturn	1320
Uranus	2600
Neptune	5160

FIGURE 1.7 Hinrichs's table of planetary distances (1864). G. D. Hinrichs, *American Journal of Science and Arts*, 2, 37, 36–56, 1864.

Hinrichs expresses the differences in these distances by the formula $2^x \times n$, in which *n* is the difference in the distances of Venus and Mercury from the sun, or 20 units. Depending on the value of *x*, the formula gives the following distances:[18]

$2^0 \times 20 = 20$
$2^1 \times 20 = 40$
$2^2 \times 20 = 80$
$2^3 \times 20 = 160$
$2^4 \times 20 = 320$
and so on.

A few years previously, in 1859, Kirchhoff and Bunsen, in Germany, had discovered that each element emitted light, which could then be dispersed with a glass prism and analyzed quantitatively.[19] They had also discovered that every element gave a unique spectrum consisting of a set of specific lines. Whereas some authors suggested that these spectral lines might provide information about the various elements that had produced them, Bunsen in particular remained quite opposed to the idea.[20]

Hinrichs, however, has no hesitation in connecting spectra with the atoms of the elements. He becomes interested in the fact that, with any particular element, the frequencies of its spectral lines always seem to be whole number multiples of the smallest difference. For example, in the case of calcium, a ratio of 1:2:4 had been observed among its spectral frequencies. Hinrichs points out that as the sizes of planetary orbits produce a regular series of whole numbers, so the ratios among spectral line differences also produce whole number ratios. He claims that the cause of the latter ratios might be due to size ratios among the atomic dimensions of the various elements (fig. 1.8).

This idea leads Hinrichs to a successful, and novel, means of classifying the elements into a periodic system. By closely studying the work of Kirchhoff and Bunsen, Hinrichs finds that some of the spectral line frequencies can be related to the chemistry of the elements through their atomic weights, as well as to their postulated "atomic dimensions" (fig. 1.9). The difference between the spectral line frequencies seem to be inversely proportional to the atomic weights of the elements in question. Hinrichs quotes the values of Ca, where the frequency difference is 4.8 units, and barium, which is chemically similar but has a higher atomic weight and shows a frequency difference of 4.4 units.[21]

The culmination of Hinrichs's work on the classification of the elements is his spiral periodic system as shown in fig. 1.10. The eleven "spokes" radiating from the center of this wheel-like system consist of three predominantly nonmetal groups and eight groups containing metals. From a modern perspective the nonmetal

From astronomy	Size ratios among orbits	⇨	whole number ratios
From spectra	Observation of whole number ratios	⇨	size ratios among atomic dimensions

FIGURE 1.8 Schematic form of Hinrichs's argument.

oxygen group;	quadratic formula			$A = n.4^2$	
	n	A	calc.	det.	Difference
oxygen	1	1.4^2	= 16	16	0.0
sulfur	2	2.4^2	= 32	32	0.0
selenium	5	5.4^2	= 80	80	0.0
tellurium	8	8.4^2	= 128	128	0.0
alkali metal group;	quadratic with pyramid			$A = 7 + n.4^2$	
lithium	0	7		7	0.0
sodium	1	$7 + 1.4^2$	= 23	23	0.0
potassium	2	$7 + 2.4^2$	= 39	39	0.0
rubidium	5	$7 + 5.4^2$	= 87	85.4	– 1.6
cesium	8	$7 + 8.4^2$	= 135	133	– 2.0
chlorine group;	quadratic formula			$A = n.3^2 \pm 1$	
fluorine	2	$2.3^2 + 1$	= 19	19	0.0
chlorine	4	$4.3^2 - 1$	= 35	35.5	+ 0.5
bromine	9	$9.3^2 - 1$	= 80	80	0.0
iodine	14	$14.3^2 + 1$	= 127	127	0.0
alkaline earth group;	quadratic formula			$A = n.2^2$	
magnesium	3	3.2^2	= 12	12	0.0
calcium	5	5.2^2	= 20	20	0.0
strontium	11	11.2^2	= 44	43.8	–0.2
barium	17	17.2^2	= 68	68.5	+ 0.5

FIGURE 1.9 Hinrichs's table of atomic weights and atomic dimensions for several groups of elements. Redrawn from G. D. Hinrichs, *American Journal of Science and Arts*, 2, 42, 350–368, 1866.

groups appear to be incorrectly ordered, in that the sequence is groups VI, V, and then VII when proceeding from left to right on the top of the spiral. The group containing carbon and silicon is classed with the metallic groups by Hinrichs, presumably because it also includes the metals nickel, palladium, and platinum.[22]

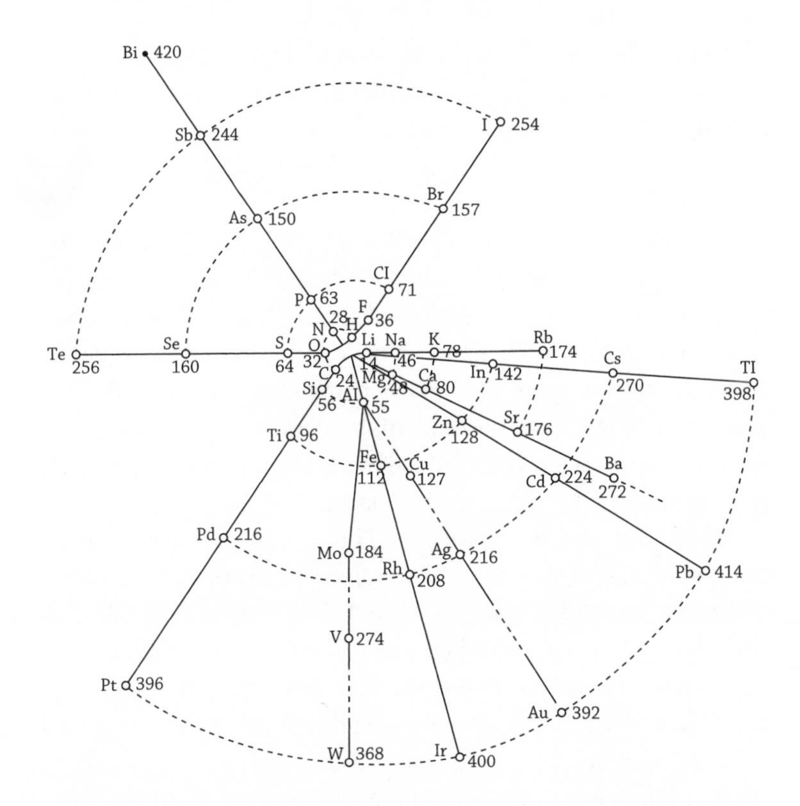

FIGURE 1.10 Hinrichs's spiral periodic system. G. D. Hinrichs, *Programm der Atomechanik oder die Chemie eine Mechanik de Pantome*, Augustus Hageboek, Iowa City, IA, 1867.

Overall, however, Hinrichs's periodic system is rather successful in grouping together many important elements. One of its main advantages is the clarity of its groupings, as compared, say, with the more elaborate but less successful periodic systems of Newlands in 1864 and 1865.[23]

It is clear from his books that Hinrichs possessed a deep knowledge of chemistry, as well as a proficiency in mineralogy.[24] Yet his approach to the classification of the elements was only partly chemical. He was, perhaps, the most interdisciplinary of all the discoverers of the periodic system. Indeed, the fact that Hinrichs arrived at his system from such a different direction as the others might be taken to lend the periodic system itself independent support, just as do Lothar Meyer's studies of physical periodicity, which will shortly be considered.

In an article published in *The Pharmacist*, in 1869, Hinrichs discusses previous unsuccessful attempts to classify the elements but fails to mention any of his codiscoverers such as de Chancourtois, Newlands, Odling, Lothar Meyer, or Mendeleev. Indeed, Hinrichs characteristically appears to have completely ignored all other attempts to base the classification of the elements directly on atomic weights, even though one can assume that he was aware of them, given his knowledge of foreign languages.[25]

Finally, it should be stressed that Hinrichs appears to be ahead of his time in assigning great importance to the analysis of the spectra of the elements and in trying to relate these facts to the periodic classification.[26] As we will see when we begin to discuss the seven missing elements discovered during the twentieth century, spectroscopy becomes, perhaps, the primary method for identifying new elements.

More than that of any other scientist discussed in this book, the work of Hinrichs is so idiosyncratic and labyrinthine that a more complete study of his output will be required before it can be adequately evaluated.

Julius Lothar Meyer

Julius Lothar Meyer is born in 1830 in Heilbronn, Germany. He is the fourth of seven children of a physician father and a mother whose own father is also a local physician. Julius and one of his brothers, Oskar, begin their studies with the intention of continuing this family medical tradition, but it is not long before both of them have turned to other fields. Oskar becomes a physicist, while Julius becomes one of the most influential chemists of his time.

Lothar Meyer is best remembered for his independent discovery of the periodic system, although more credit is invariably accorded to Mendeleev. The two chemists eventually become engaged in a rather bitter priority dispute, which Mendeleev apparently wins, partly due to his more forceful personality. Certainly, Mendeleev had a more complete system and went on to make predictions on the basis of his system. He was also to champion the cause of the periodic law to a far greater extent than did Lothar Meyer. But if one asks the question of who arrived at the mature periodic system first, a strong case can be made that in many crucial details, the system of Lothar Meyer was not only first but also more correct.

Lothar Meyer attends the Karlsruhe conference in 1860 and learns firsthand of Cannizzaro's groundbreaking work on the atomic weights of the elements.[27] He then edits a version of Cannizzaro's article that appears in Germany in Ostwald's series under the title of Klassiker der Wissenschaften. Lothar Meyer later described the effect that Cannizzaro's article had on him by saying, "the scales fell from my eyes and my doubts disappeared and were replaced by a feeling of quiet certainty."[28] In 1864, Lothar Meyer publishes the first edition of a chemistry textbook, *Die Modernen Theorie der Chemie,* which was deeply influenced by the work of Cannizzaro. The book appears in five editions and is translated into English, French, and Russian, eventually becoming one of the most authoritative treatments on the theoretical principles of chemistry before the advent of physical chemistry in the late 1800s.

By the time Lothar Meyer has written the manuscript for his book in 1862, he has produced a table of twenty-eight elements arranged in order of increasing atomic weight. An adjacent table containing a further twenty-two elements also appeared in the book, although these are not arranged according to atomic weight order. All this took place only two years after the Karlsruhe meeting.[29]

Lothar Meyer publishes his table of twenty-eight elements for the first time in 1864 (fig. 1.11). His arrangement of elements in order of increasing atomic weights and the clear establishment of

	4 werthig	3 werthig	2 werthig	1 werthig	1 werthig	2 werthig
	—	—	—	—	Li = 7.03	(Be = 9.3?)
Differenz =	—	—	—	—	16.02	(14.7)
	C = 12.0	N = 14.04	O = 16.00	Fl = 19.0	Na = 23.05	Mg = 24.0
Differenz =	16.5	16.96	16.07	16.46	16.08	16.0
	Si = 28.5	P = 31.0	S = 32.07	Cl = 35.46	K = 39.13	Ca = 40.0
Differenz =	$\frac{89.1}{2} = 44.55$	44.0	46.7	44.51	46.3	47.6
	—	As = 75.0	Se = 78.8	Br = 79.97	Rb = 85.4	Sr = 87.6
Differenz =	$\frac{89.1}{2} = 44.55$	45.6	49.5	46.8	47.6	49.5
	Sn = 117.6	Sb = 120.6	Te = 128.3	I = 126.8	Cs = 133.0	Ba = 137.1
Differenz =	89.4 = 2 × 44.7	87.4 = 2 × 43.7	—	—	(71 = 2 × 35.5)	—
	Pb = 207.0	Bi = 208.0	—	—	(Tl = 204?)	—

FIGURE 1.11 Lothar Meyer's periodic system of 1864. J. Lothar Meyer, *Die Moderne Theoriene der Chemie,* Breslau, Wroklaw, 1864. English translation, Longmans, London, 1888.

horizontal relationships among these elements is another instance in which Lothar Meyer anticipates Mendeleev by several years.[30]

Lothar Meyer's 1864 table also shows clearly for the first time a regular variation in valency of the elements, from 4 to 1 on moving from left to right across the table, followed by a repetition of the valency 1 and a further increase to elements with valency of 2.[31] This table suggests that Lothar Meyer has struggled to arrange elements in terms of atomic weight as well as chemical properties. He seems to have decided to let chemical properties outweigh strict atomic weight ordering in some cases. An example of this is in his grouping of tellurium with elements such as oxygen and sulfur, while iodine (symbol J for jod) is grouped with the halogens, in spite of its lower atomic weight. Lothar Meyer also separates the elements into two tables in a manner corresponding to the separation of our modern main-group elements from the modern transition elements.[32]

Another very noteworthy feature of Lothar Meyer's table of 1862 (published in 1864) is the presence of many gaps to denote unknown elements. Once again it appears that the leaving of gaps did not originate with Mendeleev, who was to wait a further five years before even venturing to publish a periodic system and eventually making the detailed predictions for which he subsequently became so well known.

Lothar Meyer's table contains interpolations between neighboring elements. In the space below the element silicon, for example, he indicates that there should be an element whose atomic weight would be greater than silicon's by a difference of 44.55. This implies an atomic weight of 73.1 for this unknown element, which when discovered was found to have an atomic weight of 72.3. This prediction of the element germanium, which was first isolated in 1886, is usually attributed to Mendeleev even though it is clearly anticipated by Lothar Meyer in this early table of 1864.[33]

But perhaps Lothar Meyer's greatest strength lay in his additional knowledge of physical properties and his use of them in constructing representations of the periodic system. He paid close attention to atomic volumes, densities, and fusibilities of the elements, for example. His published diagram showing the periodicity among atomic volumes of the elements (i.e., atomic weight divided by specific gravity) did much to encourage the general acceptance of the periodic system (fig. 1.12), since one can see the periodicity among the elements almost at a glance.

In the course of the controversy between Mendeleev and Lothar Meyer, which followed the publication of their respective periodic systems, Mendeleev appears to have been the victor. However, there is a rather intriguing episode, which did not come to light until much later. Had it become known earlier it might well have made a significant difference. When Lothar Meyer was preparing the second edition of his book in 1868, he produced a vastly expanded

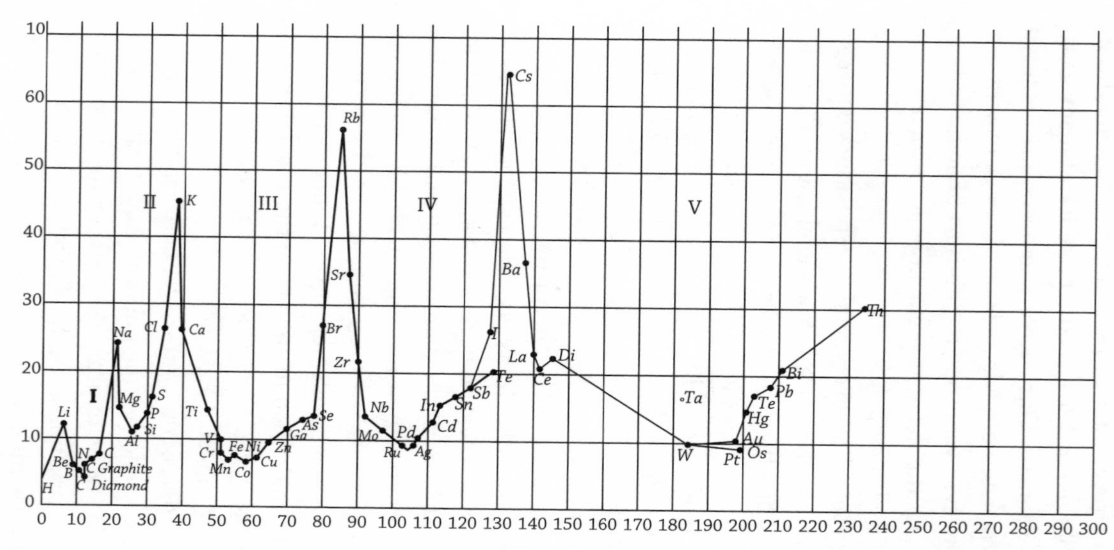

FIGURE 1.12 Lothar Meyer's atomic volume graph. J. Lothar Meyer: Die Natur der chemischen Elemente als Funktion ihrer Atomgewichte, *Lieb. Ann. Suppl.* 7, 354–364, 1870.

periodic system that included a further twenty-four elements and nine new vertical families of elements (fig. 1.13). This system preceded Mendeleev's famous table of 1869 that has subsequently been regarded as *the* mature periodic system. Moreover, Lothar Meyer's system was more accurate than Mendeleev's. For example, Lothar Meyer correctly placed mercury with cadmium, lead with tin, and thallium with boron, whereas Mendeleev's table failed to make any of these connections.[34]

It appears that for some inexplicable reason Lothar Meyer's 1868 table was not published. No less than twenty-five years later another German chemist, Adolf Remelé, showed the table to Lothar Meyer, who by this time had forgotten all about its existence. In 1895, after

MEYER'S TABLE OF 1868.

1	2	3	4	5	6	7	8
		Al=27.3 28.7/2=14.8	Al=27.3				C=12.00 16.5
							Si=28.5 89.1/2=44.5
Cr=52.6	Mn=55.1 49.2	Fe=56.0 48.9	Co=58.7 47.8	Ni=58.7	Cu=63.5 44.4	Zn=65.0 46.9	89.1/2=44.5
	Ru=104.3 92.8=2.46.4	Rh=103.4 92.8=2.46.4	Pd=106.0 93=2.465		Ag=107.9 88.8=2.44.4	Cd=111.9 88.3=2.44.5	Sn=117.6 89.4=2.41.7
	Pt=197.1	Ir=197.1	Os=199.		Au=196.7	Hg=200.2	Pb=207.0
9	10	11	12	13	14	15	
			Li=7.03 16.02	Be=9.3 14.7			
N=14.4 16.96	O=16.00 16.07	F=19.0 16.46	Na=23.05 16.08	Mg=24.0 16.0			
P=31.0 44.0	S=32.07 46.7	Cl=35.46 44.5	K=39.13 46.3	Ca=40.0 47.6	Ti=48 42.0	Mo.=92.0 45.0	
As=75.0 45.6	Se=78.8 49.5	Br=79.9 46.8	Rb=85.4 47.6	Sr=87.6 49.5	Zr=90.0 47.6	Vd=137.0 47.0	
Sb=120.6 87.4=2.43.7	Te=128.3	I=126.8	Cs=133.0 71=2.35.5	Ba=137.1	Ta=137.6	W=184.0	
Bi=208.0			Te=204.0				

FIGURE 1.13 Lothar Meyer's system of 1868. Unpublished system.

Lothar Meyer's death, Seubert, another of his colleagues, finally published the forgotten table. This attempt, following such a long time delay, failed to restore much sense of priority to Lothar Meyer.

Dmitri Mendeleev

Dmitri Ivanovich Mendeleev is the most famous Russian scientist of the modern era. He is the main discoverer of the periodic system but also recognized that this system pointed to a deep law of nature, the periodic law. He then spent several years drawing out the full consequences of this law, most notably by predicting the existence and properties of numerous new elements. He also corrected the atomic weights of some already known elements and successfully changed the position of some other elements in the periodic table.

Most important of all, Mendeleev made the periodic table his own, by pursuing its study and development during several periods during his life, even though he worked in a number of other very diverse fields. By contrast, most of his precursors or codiscoverers failed to follow up their initial discoveries. As a result, the name of Mendeleev is invariably linked with the periodic table in much the same way that evolution by natural selection and relativity theory are linked with Darwin and Einstein, respectively.

When Mendeleev was a young boy, his father went blind and died soon afterward. Dmitri, the youngest of fourteen siblings, was brought up by his adoring mother, who was determined to provide the best possible education for him. She took the young Mendeleev on a trip of hundreds of miles in an attempt to enroll him into Moscow University. But Mendeleev was rejected because of his Siberian origins and the fact that the university only admitted Russians. Undeterred, Mendeleev's mother succeeded in enrolling him into the Main Pedagogical Institute of St. Petersburg, where he began to study chemistry, physics, and biology. Unfortunately, Mendeleev's mother died very soon after he had entered this institution.

After completing his undergraduate education Mendeleev spent some time in France, and later Germany, where he was attached to the laboratory of Robert Bunsen, although he preferred to stay at home and conduct his own experiments on the nature of gases. During this period Mendeleev was fortunate enough to attend the Karlsruhe conference of 1860, not because he was a prominent chemist but more because he happened to be at the right place at the right time. It was a pivotal conference at which the leading European scientists presented their views on atomic weights and the nature of atoms and molecules. Mendeleev quickly grasped the value of Cannizzaro's ideas, as did Lothar Meyer, although Mendeleev's conversion to using Cannizzaro's atomic weights appears to have taken considerably more time than did Lothar Meyer's.

In 1865, Mendeleev defended his doctoral thesis on the interaction between alcohol and water and began to work on a book on inorganic chemistry in order to improve the teaching of this subject. In the first volume of this new book he treated the more common elements in no particular order. By 1868, he had completed this volume and began to consider how he should make the transition to the remaining elements in a second volume.

Mendeleev's Discovery of the Periodic System

Although Mendeleev had been thinking about elements, atomic weights, and classification for about ten years, he appears to have had his eureka moment on February 17, 1869. On this day he cancelled a consultancy trip and decided to work on what was to become his periodic table.

First, he listed the symbols for a handful of elements in two rows:

Na	K	Rb	Cs
Be	Mg	Zn	Cd

Then, he produced a larger array of sixteen elements.

F	Cl	Br	I				
Na	K	Rb	Cs			Cu	Ag
Mg	Ca	Sr	Ba	Zn	Cd		

By that evening Mendeleev had sketched out a periodic table, which included sixty-three known elements. He also included several gaps for unknown elements and even predicted atomic weights for some of these elements. Two hundred copies of this first table were printed and sent to chemists all over Europe. On March 6th of the same year the discovery was announced at a meeting of the newly formed Russian Chemical Society. Within a month an article had appeared in the journal of the newly formed society and another longer article appeared in Germany.

Mendeleev's scientific approach differed considerably from that of his rival, Lothar Meyer, in that Mendeleev did not believe in the unity of all matter and did not support Prout's hypothesis of the composite nature of all elements. Mendeleev also took care to distance himself from the notion of triads of elements. For example, he proposed that the element fluorine should be grouped together with chlorine, bromine, and iodine even though this would imply going beyond a triad to form a group of at least four elements (fig. 1.14).

Whereas Lothar Meyer had concentrated on physical principles and mainly on the physical properties of the elements, Mendeleev concentrated on their chemical properties. But when it came to deciding upon the most important criterion for classifying the elements, Mendeleev insisted that atomic weight ordering would tolerate no exceptions. While Mendeleev's precursors had recognized the importance of atomic weight ordering, Mendeleev reached a deeper philosophical understanding of atomic weights and the nature of elements, which allowed him to move into unchartered territory.

MENDELÉEFF'S TABLE I.—1871.

Series.	Group I. R_2O.	Group II. RO.	Group III. R_2O_3.	Group IV. RH_4. RO_2.	Group V. RH_3. R_2O_5.	Group VI. RH_2. RO_3.	Group VII. RH. R_2O_7.	Group VIII. RO_4.
1	H=1							
2	Li=7	Be=9.4	B=11	C=12	N=14	O=16	F=19	
3	Na=23	Mg=24	Al=27.3	Si=28	P=31	S=32	Cl=35.5	
4	K=39	Ca=40	—=44	Ti=48	V=51	Cr=52	Mn=55	Fe=56, Ce=59 Ni=59, Cu=63
5	(Cu=63)		—=68	—=72	As=75	Se=78	Br=80	
6	Rb=85	Sr=87	? Y=88	Zr=90	Nb=94	Mo=96	—=100	Ru=194, Rh=104 Pd=106, Ag=108
7	(Ag=108)	Cd=112	In=113	Sn=118	Sb=122	Te=125	I=127	
8	Cs=133	Ba=137	? Di=138	? Ce=140	....	....	....	....
9	....	....	....	....	....	....	....	....
10	....	....	? Er=178	? La=180	Ta=182	W=184	....	Os=195, In=197 Pt=198, Au=199
11	(Au=199)	Hg=200	TI=204	Pb=207	Bi==208	....	....	
12	....	....	....	Th=231	....	U=240	....	....

FIGURE 1.14 Mendeleev's periodic table of 1871. D. Mendeleev, *Zhurnal Russkoe Fiziko-Khimicheskoe Obshchestvo* 3, 25 (1871).

Mendeleev on the Nature of Elements

There is a long-standing puzzle in chemistry. When sodium and chlorine combine, for example, they form a completely new substance, sodium chloride. But the combining elements do not seem to survive in a compound, at least from a visual perspective. This is the phenomenon of chemical bonding, which differs markedly from the physical mixing of powdered sulfur and iron filings, for example.

The issue is to understand how, if at all, the combining elements survive in the compound. This matter is further complicated by the fact that in several languages, such as English, we still use the word "element" to refer to the combined substance—such as chlorine when it is present in sodium chloride. Moreover, what underlies both the uncombined green gas chlorine and combined chlorine is sometimes also called "element." We, therefore, have three senses of the same central chemical term to describe the substances that the periodic table is supposed to classify.

The third sense of "element," as mentioned above, has been called the metaphysical element, abstract element, transcendental element, and more recently, "element as a basic substance." This is the notion of an element as an abstract bearer of properties, but lacking such phenomenal properties as chlorine's green color. Green chlorine, meanwhile, is said to be the element existing as a "simple substance."[35]

When Antoine Lavoisier revolutionized chemistry at the end of the eighteenth century, one of his contributions was to concentrate attention onto the element as a simple substance—that is, the element in isolated form. This move was intended to improve chemistry by ridding it of excess metaphysical baggage and it was indeed a great step forward. According to Lavoisier, an "element" was to be regarded primarily as the final step in the separation of the components in any compound. Whether Lavoisier intended to banish the more abstract and philosophical sense of "element" is a debated issue, but certainly this latter sense began to assume a less important role than that of an element that could be isolated.

Nonetheless, the more abstract sense was not completely forgotten and Mendeleev was one of the chemists who proposed to elevate its status. In fact, he repeatedly claimed that the periodic system was primarily a classification of the more abstract sense of the term element and not necessarily the more concrete element that could be isolated.

When he was armed with this notion, Mendeleev was able to take a deeper view of elements than could chemists who only restricted themselves to elements in isolated form. This deeper philosophical view gave Mendeleev the possibility of going beyond appearances. If any element did not appear to fit within a particular group, Mendeleev could draw upon the deeper sense of the term "element" and thus to some extent could ignore the apparent properties of the element as an isolated or simple substance.

Predictions

One of Mendeleev's greatest triumphs, and perhaps the one that he is best remembered for, is the correct prediction of the existence of several new elements. In addition, Mendeleev adjusted the atomic weights of some elements, as well as relocating others to new positions within the periodic table. As suggested in the previous section, such farsightedness may have been due to a greater philosophical understanding of the nature of the elements than his competitors possessed. By concentrating on the more abstract concept of elements, Mendeleev was able to overcome any apparent obstacles that arose from taking the properties of the isolated elements at face value.

Although Mendeleev placed the greatest importance on the atomic weights of the elements, he also considered chemical and physical properties and the family resemblances among them. While Lothar Meyer concentrated on physical properties, Mendeleev paid more attention to the chemistry of the elements. Another criterion

Mendeleev used was that each element should occupy a single place in the periodic table, although he was willing to violate this notion when it came to what he called group VIII (fig. 1.14), in which as many as four elements occurred in one place. The more important criterion, however, was the ordering of the elements according to increasing atomic weights. In one or two cases he appeared to even violate this principle, although a closer inspection shows that the issue is more complicated.

The case of tellurium and iodine is one of only five pair reversals in the periodic system. These are consecutive elements that appear in the reverse order than they might according to increasing values of atomic weight (Te = 127.6, I = 126.9, and yet Te appears before I). Many accounts tell of Mendeleev's wisdom of reversing the positions of these elements, thus putting chemical properties over and above the ordering according to atomic weight. Such a claim is incorrect, however. First of all, Mendeleev was not the first chemist to make this particular reversal. Odling, Newlands, and Lothar Meyer all published tables in which the positions of tellurium and iodine had been reversed, well before the appearance of Mendeleev's articles. Second, Mendeleev was not placing a greater emphasis on chemical properties than on atomic weight ordering.

In fact, Mendeleev maintained the criterion of ordering the elements according to increasing atomic weight and repeatedly stated that this principle would tolerate no exceptions. Instead, he believed that the atomic weights for one, or both, of these elements had been incorrectly determined, and that future work would reveal that, even on the basis of atomic weight ordering, tellurium should be placed before iodine. Unfortunately, Mendeleev was simply wrong in maintaining this view.

At the time when Mendeleev proposed his first periodic systems, the atomic weights for tellurium and iodine were thought to be 128 and 127, respectively. Mendeleev's belief that atomic weight was the fundamental ordering principle left him with no choice but to question the accuracy of these values. This was because it was clear that in terms of chemical properties, tellurium should be grouped with the elements in group VI and iodine with those in group VII. He therefore claimed that this pair of elements had to be "reversed."

Initially, Mendeleev doubted the atomic weight of tellurium while believing that that of iodine was correct. He began to list tellurium as having an atomic weight of 125 in some of his periodic tables. At one time, Mendeleev claimed that the commonly reported value of 128 was the result of measurements made on a mixture of tellurium and a new element that he called eka-tellurium. Partly motivated by these claims, the Czech chemist Bohuslav Brauner undertook a series of experiments aimed at the redetermination of the atomic weight of tellurium. By 1883, Brauner reported that the value for tellurium should be changed to 125. In 1889, Brauner obtained new results that seemed to further strengthen the earlier finding of Te = 125. Naturally, Mendeleev was pleased by these findings.

In 1895, however, Brauner began reporting a new value for tellurium that was greater than that of iodine, thus returning matters to their initial starting point. Mendeleev responded by now starting to question the accuracy of the accepted atomic weight for iodine instead of tellurium. He requested a redetermination for the atomic weight of iodine and hoped that its value would turn out to be higher. In some of his later periodic tables Mendeleev even listed tellurium and iodine as both having atomic weights of 127. The problem would not be resolved until 1913–1914 by Moseley, who showed that the elements should be ordered according to atomic number rather than atomic weight. While tellurium has a higher atomic weight than iodine, it has a lower atomic number, and this is why it should rightly be placed before iodine in full agreement with its chemical behavior.

Mendeleev's Predictions

Although Mendeleev's predictions might seem miraculous, they were based on careful interpolation between the properties of elements flanking the unknown elements. He already began to make predictions in his first publication on the periodic system in 1869, although he published a more detailed account in a long paper of 1871. He first focused on two gaps, one below aluminium and the other below silicon, which he provisionally called eka-aluminium and eka-boron, in which the Sanskrit prefix meant "one-like." In his paper of 1869 Mendeleev wrote,

> We must expect the discovery of yet unknown elements, e.g. elements analogous to Al and Si, with atomic weights 65–75.

In 1870, he began to make predictions about a third element, which would lie below boron in the periodic table. The next year he predicted their atomic weights to be

eka-boron	eka-aluminium	eka-silicon
44	68	72

and gave detailed predictions on various chemical and physical properties of each element. It took six years before the first of these predicted elements, later called gallium, was isolated. With a few very minor exceptions, Mendeleev's predictions were almost exactly

Property	Eka-silicon 1871 prediction	Germanium discovered 1886
Relative atomic mass	72	72.32
Specific gravity	5.5	5.47
Specific heat	0.073	0.076
Atomic volume	13 cm^3	13.22 cm^3
Color	dark gray	grayish-white
Specifc gravity of dioxide	4.7	4.703
Boiling point of tetrachloride	100°C	86°C
Specific gravity of tetrachloride	1.9	1.887
Boiling point of tetra ethyl deriv.	160°C	160°C

FIGURE 1.15 The predicted and observed properties of eka-silicon (germanium).

correct. The accuracy of Mendeleev's predictions can also be clearly seen in the case of the element he called eka-silicon, later called germanium, after it had been isolated by the German chemist Clemens Winkler (fig. 1.15).

Mendeleev's Less Successful Predictions

Not all of Mendeleev's predictions were as successful as those of gallium, germanium, and scandium. As fig. 1.16 shows, he was unsuccessful in as many as nine out of his eighteen published predictions, although perhaps not all of these predictions should be given the

Element as given by Mendeleev	Predicted A.W.	Measured A.W.	Eventual name
coronium	0.4	not found	not found
ether	0.17	not found	not found
eka-boron	44	44.6	scandium
eka-cerium	54	not found	not found
eka-aluminium	68	69.2	gallium
eka-silicon	72	72.0	germanium
eka-manganese	100	99	technetium (1939)
eka-molybdenum	140	not found	not found
eka-niobium	146	not found	not found
eka-cadmium	155	not found	not found
eka-iodine	170	not found	not found
eka-caesium	175	not found	not found
tri-manganese	190	186	rhenium (1925)
dvi-tellurium	212	210	polonium (1898)
dvi-caesium	220	223	francium (1939)
eka-tantalum	235	231	protactinium (1917)

FIGURE 1.16 Mendeleev's predictions, successful and otherwise (compiled by the author).

same weight. This is because some of the elements involved the rare earths, which resemble each other very closely and which posed a major challenge to the periodic table for many years to come.[36]

In addition, Mendeleev's failed predictions raise an interesting philosophical point. For some time, historians and philosophers of science have debated whether successful predictions should count more, or less, than the successful accommodation of already known data. On one hand, there is no disputing the fact that successful predictions carry a great deal of psychological impact, since they imply that the scientist can foretell the future. But successful accommodation, or explanation of already known data, is also rather impressive, especially since there is usually a great deal more known information to incorporate into a new scientific theory. This was the case with Mendeleev and the periodic table, since he had to successfully accommodate as many as sixty-three known elements into a fully coherent system.

At the time of discovery of the periodic table, the Nobel Prizes had not yet been instituted. One of the most prestigious prizes in chemistry was the Davy Medal, awarded by the Royal Society of Chemistry and named after the chemist Humphry Davy. In 1882, the Davy Medal was jointly awarded to Lothar Meyer and Mendeleev. This fact seems to suggest that the chemists making the award were willing to recognize Lothar Meyer, who had not made any predictions to speak of. In addition, the citation that accompanied the award to Lothar Meyer and Mendeleev made no mention whatsoever of Mendeleev's successful predictions. It would appear that this group of prominent British chemists was not persuaded by the psychological impact of successful predictions over the ability of a scientist to successfully accommodate the known elements.

The Noble Gases

The discovery of the noble gases by Ramsay and Rayleigh at the end of the nineteenth century represented an interesting challenge to the periodic table. First of all, in spite of Mendeleev's dramatic predictions of many elements, he failed to foresee this entire group of elements (He, Ne, Ar, Kr, Xe). Moreover, nobody else predicted these elements or even suspected their existence.

The first noble gas to be isolated was argon, which was obtained in small amounts in 1894 at London's University College by Lord Rayleigh and William Ramsay while working with the gas nitrogen. Unlike with many previously discussed elements, however, a number of factors seemed to conspire together to make the accommodation of this element a very difficult task. The all-important atomic weight of argon—required if it were to be placed in the periodic table—was not easily obtained. This was due to the fact that it was not clear how many atoms of the element were combined together to form its molecules. Most measurements pointed to its being monoatomic, while all the other gaseous elements at the time were diatomic (H_2, N_2, O_2, F_2, Cl_2). If argon was indeed monoatomic, its atomic weight would be approximately 40, which rendered its accommodation into the periodic table rather problematic since there was no gap at this value in the periodic table (see fig. 1.14). The element calcium has an atomic weight of about 40, followed by scandium, one of Mendeleev's successfully predicted elements, with an atomic weight of 44, thus leaving no apparent space for a new element with an atomic weight of 40.

There *was* a rather large gap between chlorine (35.5) and potassium (39), but placing argon between these two elements would have produced a rather obtrusive pair reversal. It should also be recalled that at this time only one significant such pair reversal existed, involving the elements tellurium and iodine, and this behavior was regarded as being highly anomalous. As mentioned above, Mendeleev had concluded that the Te–I pair reversal was due to incorrect atomic weight determinations on either tellurium or iodine, or perhaps both elements. The notion that argon and potassium would represent another such case was therefore regarded with a great deal of suspicion.

Another unusual aspect of the element argon was its complete chemical inertness, which meant that its compounds could not be studied for the simple reason that none existed. Some took the inertness of the gas to mean that it was not a genuine chemical element and that it would not need to be placed anywhere in the periodic table.

Many others, however, persisted in attempting to place the element into the periodic table. The accommodation of argon was the focus of a general meeting of the Royal Society in 1885. The discoverers, Rayleigh and Ramsay, argued that the element was probably monoatomic but admitted that they could not be sure. Nor could they be certain that the gas in question was not a mixture, which might imply that the atomic weight was not in fact 40. William Crookes presented some evidence in favor of the sharp boiling and melting points of argon, thus pointing to its being a single element rather than a mixture. Henry Armstrong, a leading chemist, argued that argon might behave like nitrogen, in that it would form an inert diatomic molecule even though its individual atoms might be highly reactive. A physicist, Rücker, argued that an atomic weight of approximately 40 was correct and that if this element could not be placed in the periodic table, then it was the periodic table itself that was at fault.

The findings of the Royal Society meeting were thus inconclusive about the fate of the new element or whether argon should even count as a new element. Mendeleev himself, who did not attend this meeting, published an article in the London-based *Nature* magazine in which he concluded that argon was, in fact, triatomic and that it consisted of three atoms of nitrogen. He based this notion on the fact that the assumed atomic weight of 40 divided by 3 gives approximately 13.3, which is not so far removed from 14, the atomic weight of nitrogen. In addition, argon had been discovered in the course of experiments on nitrogen gas, which rendered the triatomic idea somewhat plausible.

The issue was finally resolved in the year 1900. Ramsay, one of the codiscoverers of the new gas, wrote to Mendeleev informing him that the new group, which had by then been augmented with helium, neon, krypton, and xenon, could be easily accommodated in an eighth column between the halogens and the alkali metals. Argon, the first of these new elements, had been especially troublesome because it represented a new case of a genuine pair reversal. It has an atomic weight of about 40 and yet appears before potassium, with an atomic weight of about 39. Mendeleev readily accepted this proposal and later wrote,

> This was extremely important for him [Ramsay] as an affirmation of the position of the newly discovered elements, and for me as a glorious confirmation of the general applicability of the periodic law.

Rather than threatening the periodic table, the discovery of the noble gases and their successful accommodation into the periodic table served to strengthen the power and generality of Mendeleev's periodic system.

Chapter 2

The Invasion of the Periodic Table by Physics

Although John Dalton had reintroduced the notion of atoms to science, many debates followed among chemists, most of whom refused to accept that atoms existed literally. One of these skeptical chemists was Mendeleev, but as we saw in the previous chapter this does not seem to have prevented him from publishing the most successful periodic system of all those proposed at the time. Following the work of physicists like Einstein and Perrin, the atom's reality became more and more firmly established starting at the turn of the twentieth century. Einstein's 1905 paper on Brownian motion, using statistical methods, provided conclusive theoretical justification for the existence of atoms but lacked experimental support.[1] The latter was soon provided by the French experimental physicist Jean Perrin.[2]

This work led in turn to many lines of research aimed at exploring the structure of the atom, and many developments that were to have a big influence on attempts to understand the periodic system theoretically. In this chapter we consider some of this atomic research as well as several other key discoveries in twentieth-century physics that contributed to what might be called the invasion of the periodic table by physics.

The Electron and Other Discoveries in Physics

The discovery of the electron, the first hint that the atom had a substructure, came in 1897 at the hands of the legendary J. J. Thomson, working at the Cavendish laboratory in Cambridge. A little earlier, in

1895, Wilhelm Conrad Röntgen had discovered X-rays in Würzburg, Germany. These new rays would soon be put to very good use by Henry Moseley, a young physicist working first in Manchester and, for the remainder of his short scientific life, in Oxford.

Just a year after Röntgen had described his X-rays, Henri Becquerel in Paris discovered the enormously important phenomenon of radioactivity, whereby certain atoms break up spontaneously while emitting a number of different, new kinds of rays. The term "radioactivity" was actually coined by the Polish-born Marie Slodowska (later Curie). With her husband, Pierre Curie, she took up the work on this dangerous new phenomenon and soon discovered a couple of new elements that they called polonium and radium.

By studying how atoms break up while undergoing radioactive decay, it became possible to probe the components of the atom more effectively, as well as the laws that govern how atoms transform themselves into other atoms. So although the periodic table deals with atoms of different elements, there seems to be features that allow some atoms to be converted into others under some conditions. For example, the loss of an α particle, consisting of a helium nucleus with two protons, results in a lowering of the atomic number of an element by two units.

Another very influential physicist working at this time was Ernest Rutherford, a New Zealander who arrived in Cambridge to pursue a research fellowship, later spending periods of time at McGill University in Montreal and Manchester University in England. He then returned to Cambridge to assume the directorship of the Cavendish Laboratory as the successor to J. J. Thomson. Rutherford's contributions to atomic physics were many and varied and included the discovery of the laws governing radioactive decay as well as his "splitting of the atom." He was also the first to achieve the "transmutation" of elements into other new elements. In this way Rutherford achieved an artificial analogue to the process of radioactivity, which similarly yielded atoms of a completely different element and once again emphasized the essential unity of all forms of matter.[3]

Another discovery by Rutherford consisted of the nuclear model of the atom, a concept that is taken more or less for granted these days, namely the notion that the atom consists of a central nucleus surrounded by orbiting negatively charged electrons.[4]

Nevertheless, Rutherford was not the first to suggest a nuclear model of the atom, which resembled a miniature solar system. That distinction belongs to the French physicist Jean Perrin, who in 1900 proposed that negative electrons circulated around the positive nucleus like planets circulating around the sun. In 1903, this astronomical analogy was given a new twist by Hantaro Nagaoka, who proposed a Saturnian model in which the electrons now took the place of the rings around the planet Saturn. But neither Perrin nor Nagaoka could appeal to any solid experimental evidence to support their models of the atom.

Rutherford, along with his junior colleagues Geiger and Marsden, fired a stream of α particles at a thin metal foil made of gold and obtained a very surprising result. Whereas most of the α particles passed through the gold foil in a more or less unimpeded fashion, a significant number of them were deflected at very oblique angles. Rutherford's conclusion was that atoms of gold, or anything else for that matter, consist mostly of empty space except for a dense central nucleus. The fact that some of the α particles were unexpectedly bouncing back toward the incoming stream of α particles was thus evidence for the presence of a tiny central nucleus in every atom.

Nature, therefore, turned out to be more fluid than had previously been believed. Mendeleev, for example, had thought that elements were strictly individual. He could not accept the notion that elements could be converted into different ones. After the Curies began to report experiments that suggested the breaking up of atoms, Mendeleev traveled to Paris to see the evidence for himself, close to the end of his life. It is not clear whether he accepted this radical new notion even after his visit to the Curie laboratory.

X-rays

In 1895, the German physicist Röntgen made a momentous discovery at the age of forty. Up to this point his research output had been somewhat unexceptional. As the atomic physicist Emilio Segrè wrote a good deal later,

> By the beginning of 1895 Röntgen had written forty-eight papers now practically forgotten. With his forty-ninth he struck gold.

Röntgen was in the process of exploring the action of an electric current in an evacuated glass tube called a Crookes tube. Röntgen noticed that an object on the other side of his lab, which was not part of his experiment, was glowing. He quickly established that the glow was not due to the action of the electric current and deduced that some new form of rays might be produced inside his Crookes tube. Soon Röntgen also discovered the property that X-rays are best known for. He found that he could produce an image of his hand that clearly showed the outlines of just his bones. A powerful new technique that was to provide numerous medical applications had come to light. After working secretly for seven weeks, Röntgen was ready to announce his results to the Würzburg Physical-Medical Society, an interesting coincidence given the impact that his new rays were to have on both of these two fields.

Some of Röntgen's original X-ray images were sent to Paris, where they reached Henri Becquerel, who became interested in examining the relationship between X-rays and the property of phosphorescence, whereby certain substances emit light on exposure to sunlight. In order to test this notion, Becquerel wrapped some crystals of a uranium salt in some thick paper and, because of a lack of sunlight, decided to put these materials away into a drawer for a few days. By another stroke of luck, Becquerel happened to place his wrapped crystals on top of an undeveloped photographic plate before closing the drawer and going about his business for a few more cloudy Parisian days.

On finally opening the drawer he was amazed to find an image that had formed on the photographic plate by the uranium crystals even though no sunlight had struck them. This clearly suggested that the uranium salt was emitting its own rays, irrespective of the process of phosphorescence. Becquerel had discovered nothing less than radioactivity, a natural process in some materials whereby the spontaneous decay of atoms produces powerful and dangerous emanations. It was Marie Curie who was to dub this phenomenon as "radioactivity" a few years later.

The supposed connection between these experiments and X-rays turned out to be incorrect. Becquerel failed to find any connection between X-rays and phosphorescence. In fact, X-rays had not even entered into these experiments, although he had discovered a phenomenon that was to have enormous importance in more ways than one. First, radioactivity was an early and significant step into the exploration of matter and radiation, and second, it was to lead indirectly to the development of nuclear weapons and nuclear energy.

Back to Rutherford

Around the year 1911, Rutherford reached the conclusion that the charge on the nucleus of the atom was approximately half of the weight of the atom concerned, or $Z \approx A/2$, after analyzing the results of atomic scattering experiments. This conclusion was supported by the Oxford physicist Charles Barkla, who arrived at it via an altogether different route, using experiments with X-rays.

Meanwhile, a complete outsider to the field, the Dutch econometrician Anton van den Broek, was pondering over the possibility of modifying Mendeleev's periodic table. In 1907, he proposed a new table containing 120 elements, although many of these remained as empty spaces.[5] A good number of empty spaces were occupied by some newly discovered substances whose elemental status was still in some question. They included so called thorium emanation, uranium-X (an unknown decay product of uranium), Gd_2 (a decay product from gadolinium), and many other new species.

But the really novel feature of van den Broek's work was a proposal that all elements were composites of a particle that he named the alphon, consisting of half of a helium atom with a mass of two atomic weight units. In 1911, he published a further article in which he dropped any mention of alphons but retained the idea of elements differing by two units of atomic weight. In a twenty-line letter to London's *Nature* magazine he went a step further toward the concept of atomic number by writing,[6]

> the number of possible elements is equal to the number of possible permanent charges.

Van den Broek was thus suggesting that since nuclear charge on an atom was half of its atomic weight, and that the atomic weights of successive elements increased in stepwise fashion by two, then the nuclear charge would define the position of an element in the periodic table. In other words, each successive element in the periodic table would have a nuclear charge greater by one unit than the previous element.

A further article published in 1913 drew the attention of Niels Bohr, who cited van den Broek in his own famous trilogy paper of 1913 on the hydrogen atom and the electronic configurations of many-electron atoms.[7] In the same year, van den Broek wrote another paper that also appeared in *Nature,* this time explicitly connecting the serial number on each atom with the charge on each atom. More significantly perhaps, he disconnected this serial number from atomic weight. This landmark publication was praised by many experts in the field, including Soddy and Rutherford, all of whom had failed to see the situation as clearly as the amateur van den Broek.

Moseley

Although an amateur had confounded the experts in arriving at the concept of atomic number, he did not quite complete the task of establishing this new quantity. The person who did complete it, and who is almost invariably given the credit for the discovery of atomic number, was the English physicist Henry Moseley, who died in World War I at the age of twenty-six. His fame rests on just two papers in which he confirmed experimentally that atomic number was a better ordering principle for the elements than atomic weight.[8] This research is also important because it allowed others to determine just how many elements were still awaiting discovery between the limits of the naturally occurring elements (hydrogen and uranium), which is the main concern of the present book.

Moseley received his training at the University of Manchester as a student of Rutherford's. Moseley's experiments consisted of bouncing light off the surface of samples of various elements and recording the characteristic X-ray frequency that each one emitted. Such emissions occur because an inner electron is ejected from the atom, causing an outer electron to fill the empty space, in a process that is accompanied by the emission of X-rays. Moseley first selected fourteen elements, nine of which, titanium to zinc, formed a continuous sequence of elements in the periodic table. What he discovered was that a plot of the emitted X-ray frequency against the square of an integer representing the position of each element in the periodic table produced a straight-line graph. Here was confirmation of van den Broek's hypothesis that the elements can be ordered by means of a sequence of integers, later called atomic number, one for each element starting with H = 1, He = 2, and so on. In a second paper, he extended this relationship to an additional thirty elements, thus further solidifying its status.

It, therefore, became a relatively simple matter for Moseley to verify whether the claims for many newly claimed elements were valid or not. For example, the Japanese chemist Ogawa had claimed that he had isolated an element to fill the space below manganese in the periodic table. Moseley measured the frequency of X-rays that Ogawa's sample produced and found that it did not correspond to the value expected of element 43.

While chemists had been using atomic weights to order the elements, there had been a great deal of uncertainty about just how many elements remained to be discovered. This was due to the irregular gaps that occurred between the values of the atomic weights of successive elements in the periodic table. This complication disappeared when the switch was made to using Moseley's atomic number. Now the gaps between successive elements became perfectly regular, namely one unit of atomic number.

After Moseley died, many other chemists and physicists used his method and found that the remaining unknown elements were those with atomic numbers of 43, 61, 72, 75, 85, 87, and 91 among the evenly spaced sequence of atomic numbers. The story of their discoveries will form the remaining chapters of this book.

Isotopes

The discovery of isotopes of any particular element is another key step in understanding the periodic table that occurred at the dawn of atomic physics. The term comes from *iso* (same) and *topos* (place) and is used to describe atomic species of any particular element that differ in weight and yet occupy the same *place* in the periodic table. The discovery came about partly as a matter of necessity. The new developments in atomic physics led to the discovery of a number of new elements such as radium, radon, polonium, and actinium, which easily assumed their rightful places in the periodic table. But in addition, thirty or so apparent new elements were discovered over a short period of time. These new species were given provisional names like thorium emanation, radium emanation, actinium

X, uranium X, thorium X, and so on, to indicate the elements that seemed to be producing them.[9]

Some designers of periodic tables, such as van den Broek, attempted to accommodate these new "elements" into extended periodic tables as we saw above. Meanwhile, two Swedes, Strömholm and Svedherg, produced periodic tables in which some of these exotic new species were forced into the same place. For example, below the inert gas xenon, they placed radium emanation, actinium emanation, and thorium emanation. This seems to represent an anticipation of isotopy, but still not a very clear recognition of the phenomenon.

In 1907, the year of Mendeleev's death, the American radiochemist Herbert McCoy concluded that "radiothorium is entirely inseparable from thorium by chemical processes." This was a key observation that was soon repeated in the case of many other pairs of substances that had originally been thought to be new elements. The full appreciation of such observations was made by Frederick Soddy, another former student of Rutherford.

To Soddy the chemical inseparability meant that these were two forms, or more, of the same chemical element. In 1913, he coined the term "isotopes" to signify two or more atoms of the same element that were chemically completely inseparable but that had different atomic weights. Chemical inseparability was also observed by Paneth and von Hevesy in the case of lead and "radio lead", after Rutherford had asked them to separate them chemically. After attempting this feat by twenty different chemical approaches, they were forced to admit complete failure. Although a failure in some respects, it also solidified further the notion of one element—lead in this case—that occurs as chemically inseparable isotopes.[10]

In 1914, the case for isotopy gained even greater support from the work of T. W. Richards at Harvard, who began measuring the atomic weights of two isotopes of the same element. He too selected lead, since this element was produced by a number of radioactive decay series. The lead atoms formed by these alternative pathways, involving quite different intermediate elements, resulted in the formation of atoms of lead differing by a large value of 0.75 of an atomic weight unit.

Most significantly, the discovery of isotopes further clarified the occurrence of pair reversals, such as in the case of tellurium and iodine that had plagued Mendeleev. Tellurium has a higher atomic weight than iodine, even though it precedes it in the periodic table, because the weighted average of all the isotopes of tellurium happens to be a higher value than the weighted average of the isotopes of iodine. Atomic weight is thus a contingent quantity depending upon the relative abundance of all the isotopes of an element. The more fundamental quantity, as far as the periodic table is concerned, is atomic number or, as it was later realized, the number of protons in the nucleus. The identity of an element is captured by its atomic number and not its atomic weight, since the latter differs according to the particular sample from which the element has been isolated.

Whereas tellurium has a higher average atomic weight than iodine, its atomic number is one unit smaller. If one uses atomic number instead of atomic weight as an ordering principle for the elements, both tellurium and iodine fall into their appropriate groups in terms of chemical behavior. It, therefore, emerged that pair reversals had only been required because an incorrect ordering principle had been used in all periodic tables prior to the turn of the twentieth century.

Electronic Structure

The last section dealt mostly with discoveries in classical physics that did not require quantum theory. This was true of X-rays and radioactivity, which were largely studied without any quantum concepts, although the theory was later used to clarify certain aspects. In addition, the physics described in the last section was mostly about processes originating in the nucleus of the atom. Radioactivity is essentially about the breakup of the nucleus, and the transmutation of elements likewise takes place in the nucleus. Moreover, atomic number is a property of the nuclei of atoms, and isotopes are distinguished by different masses of atoms of the same element, which are made up almost exclusively by the mass of their nuclei.

In this section, we consider discoveries concerning the electrons in atoms, a study that did necessitate the use of quantum theory in its early days. But first we will need to say something of the origins of quantum theory itself. It all began before the turn of the twentieth century in Germany, where a number of physicists were attempting to understand the behavior of radiation held in a small cavity with blackened walls. The spectral behavior of such "black body radiation" was carefully recorded at different temperatures and attempts were then made to model the resulting graphs mathematically. The problem remained unsolved for a good deal of time, until Max Planck made the bold assumption that the energy of this radiation consisted of discrete packets or "quanta" in the year 1900. Planck himself seems to have been reluctant to accept the full significance of his new quantum theory and it was left to others to make some new applications of it.

The quantum theory asserted that energy comes in discrete bundles and that no intermediate values can occur between certain whole number multiples of the basic quantum of energy. This theory was successfully applied to the photoelectric effect in 1905 by none other than Albert Einstein, perhaps the most brilliant physicist of the twentieth century. The outcome of his research was that light could be regarded as having a quantized, or particulate, nature.[11]

In 1913, the Dane Niels Bohr applied quantum theory to the hydrogen atom that he supposed, like Rutherford, to consist of a central nucleus with a circulating electron. Bohr assumed that the energy available to the electron occurred only in certain discrete values or, in pictorial terms, that the electron could exist in any number of shells or orbits around the nucleus. This model could explain, to some extent, a couple of features of the behavior of the hydrogen atom and in fact atoms of any element. First, it explained why atoms of hydrogen that were exposed to a burst of electrical energy would result in a discontinuous spectrum in which only some rather specific frequencies were observed. Bohr reasoned that such behavior came about when an electron underwent a transition from one allowed energy level to another. Such a transition was accompanied by the release, or absorption, of the precise energy corresponding to the energy difference between the two energy levels in the atom.

Second, and less satisfactorily, the model explained why electrons did not lose energy and collapse into the nucleus of any atom, as predicted by applying classical mechanics to a charge particle undergoing circular motion. Bohr's response was that the electrons would simply not lose energy provided that they remained in their fixed orbits. He also postulated that there was a lowest energy level beyond which the electron could not undergo any downward transitions.[12]

Bohr then generalized his model to cover any many-electron atom rather than just hydrogen. He also set about trying to establish the way in which the electrons were arranged in any particular atom. Whereas the theoretical validity of making such a leap from one-electron to many-electrons was in question, this did not deter Bohr from forging ahead. The electronic configurations that he arrived at are shown in fig. 2.1.

But Bohr's assignment of electrons to shells was not carried out on mathematical grounds, nor with any explicit help from the quantum theory. Instead, Bohr appealed to chemical evidence such as the knowledge that atoms of the element boron can form three bonds, as do other elements in the boron group. An atom of boron, therefore, had to have three outer-shell electrons in order for this to be possible. But even with such a rudimentary and nondeductive theory, Bohr was providing the first successful electron-based explanation for why such elements as lithium, sodium, and potassium occur in the same group of the periodic table, and likewise for

1	H	1				
2	He	2				
3	Li	2	1			
4	Be	2	2			
5	B	2	3			
6	C	2	4			
7	N	4	3			
8	O	4	2	2		
9	F	4	4	1		
10	Ne	8	2			
11	Na	8	2	1		
12	Mg	8	2	2		
13	Al	8	2	3		
14	Si	8	2	4		
15	P	8	4	3		
16	S	8	4	2	2	
17	Cl	8	4	4	1	
18	Ar	8	8	2		
19	K	8	8	2	1	
20	Ca	8	8	2	2	
21	Sc	8	8	2	3	
22	Ti	8	8	2	4	
23	V	8	8	4	3	
24	Cr	8	8	2	2	2

FIGURE 2.1 Bohr's original 1913 scheme for electronic configurations of atoms. From, N. Bohr, On the Constitution of Atoms and Molecules, *Philosophical Magazine*, 26, 476–502, 1913, 497.

the membership to any group in the periodic table. In the case of lithium, sodium, and potassium, it is because each of these atoms has one electron that is set aside from the remaining electrons in an outer shell.

Bohr's theory had some other limitations, one of them being that it was only strictly applicable to one-electron atoms such as hydrogen or ions like He^{+}, Li^{2+}, Be^{3+}, and so on. It was also found that some of the lines for the "hydrogenic" spectra broke up into unexpected pairs of lines. Arnold Sommerfeld working in Germany suggested that the nucleus might lie at one of the foci of an ellipse rather than at the heart of a circular atom. His calculations showed that one had to introduce subshells within Bohr's main shells of electrons. Whereas Bohr's model was characterized by one quantum number denoting each of the separate shells or orbits, Sommerfeld's modified model required two quantum numbers to specify the elliptical path of the electron. Armed with the new quantum number, Bohr was able to compile a more detailed set of electronic configurations in 1923, as shown in fig 2.2.

A few years later, the English physicist Edmund Stoner found that a third quantum number was also needed to specify some finer details of the spectrum of hydrogen and other atoms. Then in 1924 the Austrian-born theorist Wolfgang Pauli discovered the need for a fourth quantum number, which was identified with the concept of an electron adopting one of two values of a special kind of angular momentum. This new kind of motion was eventually called electron "spin" even though electrons do not literally spin in the same way that the Earth spins about an axis while also performing an orbital motion around the sun.

H	1				
He	2				
Li	2	1			
Be	2	2			
B	2	3			
C	2	4			
N	2	4	1		
O	2	4	2		
F	2	4	3		
Ne	2	4	4		
Na	2	4	4	1	
Mg	2	4	4	2	
Al	2	4	4	2	1
Si	2	4	4	4	
P	2	4	4	4	1
S	2	4	4	4	2
Cl	2	4	4	4	3
Ar	2	4	4	4	4

FIGURE 2.2 Bohr's 1923 electronic configurations based on two quantum numbers. From N. Bohr, Linienspektren und Atombau, *Annalen der Physik, 71*, 228–288, 1923. p. 260.

The four quantum numbers are related to each other by a set of nested relationships. The third quantum number depends on the value of the second quantum number, which in turn depends on that of the first quantum number. Pauli's fourth quantum number is a little different since it can adopt values of +1/2 or −1/2 regardless of the values of the other three quantum numbers. The importance of the fourth quantum number, especially, is that its arrival provided a good explanation for why each electron shell can contain a certain number of electrons (2, 8, 18, 32, etc.), starting with the shell closest to the nucleus.

Here is how this scheme works. The first quantum number *n* can adopt any integral value starting with 1. The second quantum number, which is given the label ℓ, can have any of the following values related to the values of n,

$$\ell = n - 1, \ldots \ldots 0$$

In the case when n = 3, for example, ℓ can take the values 2, 1, or 0. The third quantum number, labeled m_ℓ, can adopt values related to those of the second quantum numbers as:

$$m_\ell = -\ell, -(\ell - 1), \ldots 0 \ldots (\ell - 1), \ell$$

For example, if $\ell = 2$, the possible values of m_ℓ are:

−2, −1, 0, +1, +2

Finally, the fourth quantum number, labeled m_s, can only take two possible values, either +1/2 or −1/2 units of spin angular momentum as mentioned before. There is, therefore, a hierarchy of related values for the four quantum numbers, which are used to describe any particular electron in an atom (fig. 2.3).

As a result of this scheme, it is clear why the third shell, for example, can contain a total of eighteen electrons. If the first quantum number, given by the shell number, is 3, there will be a total of $2 \times (3)^2$, or 18 electrons in the third shell. The second quantum number, ℓ,

n	*Possible values of ℓ*	*Subshell designation*	*Possible values of m_ℓ*	*Subshell*	*Electrons in each shell*
1	0	1s	0	1	2
2	0	2s	0	1	
	1	2p	1, 0, –1	3	8
3	0	3s	0	1	
	1	3p	1, 0, –1	3	
	2	3d	2, 1, 0, –1, –2	5	18
4	0	4s	0	1	
	1	4p	1, 0, –1	3	
	2	4d	2, 1, 0, –1, –2	5	
	3	4f	3, 2, 1, 0, –1, –2, –3	7	32

FIGURE 2.3 Combination of four quantum numbers to explain the total number of electrons in each shell.

can take values of 2, 1, or 0. Each of these values of ℓ will generate a number of possible values of m_ℓ and each of these values will be multiplied by a factor of two since the fourth quantum number can adopt values of 1/2 or – 1/2.

But the fact that the third shell can contain 18 electrons does not strictly explain why it is that some of the periods in the periodic system contain 18 places. It would only be a rigorous explanation of this fact if electron shells were filled in a strictly sequential manner. Although electron shells begin by filling in a sequential manner, this ceases to be the case starting with element number 19, or potassium. Configurations are built up starting with the 1s orbital, which can contain two electrons, moving to the 2s electron, which likewise is filled with another two electrons. Then come the 2p orbitals, which altogether contain a further six electrons, and so on. This process continues in a predictable manner up to element 18, or argon, which has the configuration of:

$1s^2, 2s^2, 2p^6, 3s^2, 3p^6$

One might expect that the configuration for the subsequent element, number 19, or potassium, would be

$1s^2, 2s^2, 2p^6, 3s^2, 3p^6, 3d^1$

where the final electron occupies the next subshell, which is labeled as 3d. This would be expected because up to this point the pattern has been one of adding the differentiating electron to the next available orbital at increasing distances from the nucleus. However, experimental evidence shows that the configuration of potassium should be denoted as,

$1s^2, 2s^2, 2p^6, 3s^2, 3p^6, 4s^1$

Similarly, in the case of element 20, or calcium, the new electron also enters the 4s orbital. But in the next element, number 21 or scandium, the configuration is observed to be

$1s^2, 2s^2, 2p^6, 3s^2, 3p^6, 3d^1, 4s^2$

This kind of skipping backward and forward among available orbitals as the electrons fill successive elements recurs again several times. The order of filling is often summarized in fig. 2.4, below.

As a consequence of this order of filling, successive periods in the periodic table contain the following number of elements (2, 8, 8, 18, 18, 32, etc.), thus showing a "doubling" for each period except the first one.

Whereas the rules for the combination of four quantum numbers provides a rigorous explanation for the point at which shells close, it does not provide an equally rigorous explanation for the point at which periods close. Nevertheless, some rationalizations can be given for this order of filling, although they are somewhat dependent on the facts one is trying to explain. We know where the periods close because we realize that the noble gases occur at elements 2, 10, 18, 36, 54, and so on. Similarly, we have a knowledge of the order of orbitals that must be considered in order to obtain the overall configuration, but this mnemonic rule is not derived

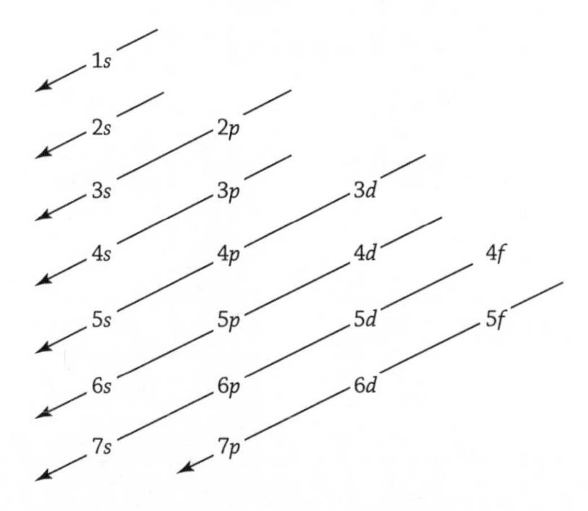

FIGURE 2.4 Sequence to follow in order to obtain the overall configuration of most atoms. Follow diagonal arrows from top to bottom.

from theory.[13] The conclusion, seldom acknowledged in textbook accounts, is that quantum physics only partly explains the periodic table. Nobody has yet deduced the Madelung rule used to predict the overall configuration from the principles of quantum mechanics. This is not to say that it might not be achieved in the future or that the order of electron filling is in any sense inherently irreducible to quantum physics. As in all branches of sciences, the process of reductionism is ongoing and provisional.[14]

After this extensive look at the background of how the periodic table emerged and how it became colonized by the field of physics, we are ready to resume the project of this book—an examination of the discovery of the last seven among the original 1–92 elements.

Chapter 3

Element 91—Protactinium

H																	He
Li	Be											B	C	N	O	F	Ne
Na	Mg											Al	Si	P	S	Cl	Ar
K	Ca	Sc	Ti	V	Cr	Mn	Fe	Co	Ni	Cu	Zn	Ga	Ge	As	Se	Br	Kr
Rb	Sr	Y	Zr	Nb	Mo	Tc	Ru	Rh	Pd	Ag	Cd	In	Sn	Sb	Te	I	Xe
Cs	Ba	Lu	Hf	Ta	W	Re	Os	Ir	Pt	Au	Hg	Tl	Pb	Bi	Po	At	Rn
Fr	Ra	Lr	Rf	Db	Sg	Bh	Hs	Mt	Ds	Rg	Cn		Fl		Lv		

La	Ce	Pr	Nd	Pm	Sm	Eu	Gd	Tb	Dy	Ho	Er	Tm	Yb
Ac	Th	**Pa**	U	Np	Pu	Am	Cm	Bk	Cf	Es	Fm	Md	No

FIGURE 3.1 A periodic table showing the position of the missing element that became known as protactinium. This table, like those that open chapters 4 to 9 inclusive, has been updated to include all officially named elements at the time of writing.

The first of our seven elements, protactinium, was one of the many elements correctly predicted by Mendeleev even in his early publications. This is not true of the famous 1896 paper, where Mendeleev used incorrect values for both thorium (118) and uranium (116). See fig 3.2.

A mere two years later, in 1871, Mendeleev corrected both of these values and indicated a missing element between thorium and uranium (fig. 1.14). But Mendeleev did not just indicate the pres-

преимущественно найдти общую систему элементовъ. Вотъ этотъ опытъ:

			Ti=50	Zr=90	?=180.
			V=51	Nb=94	Ta=182.
			Cr=52	Mo=96	W=186.
			Mn=55	Rh=104,4	Pt=197,4
			Fe=56	Ru=104,4	Ir=198.
			Ni=Co=59	Pd=106,6	Os=199.
H=1			Cu=63,4	Ag=108	Hg=200.
	Be=9,4	Mg=24	Zn=65,2	Cd=112	
	B=11	Al=27,4	?=68	Ur=116	Au=197?
	C=12	Si=28	?=70	Sn=118	
	N=14	P=31	As=75	Sb=122	Bi=210
	O=16	S=32	Se=79,4	Te=128?	
	F=19	Cl=35,5	Br=80	I=127	
Li=7	Na=23	K=39	Rb=85,4	Cs=133	Tl=204
		Ca=40	Sr=87,6	Ba=137	Pb=207.
		?=45	Ce=92		
		?Er=56	La=94		
		?Yt=60	Di=95		
		?In=75,6	Th=118?		

FIGURE 3.2 Mendeleev's earliest table, published in 1869. There is yet no sign of a prediction for eka-tantalum and the atomic values of thorium and uranium have not yet been corrected. D.I. Mendeleev, Sootnoshenie svoistv s atomnym vesom elementov, *Zhurnal Russkeo Fiziko-Khimicheskoe Obshchestv,* 1, 60–77, 1869, table on p. 70.

ence of a missing element; he added the following brief paragraph in which he ventured to make more specific predictions:

> Between thorium and uranium in this series we can further expect an element with an atomic weight of about 235. This element should form a highest oxide R_2O_5, like Nb and Ta to which it should be analogous. Perhaps in the minerals which contain these

elements a certain amount of weak acid formed from this metal will also be found.[1]

The modern atomic weight for eka-tantalum or protactinium is in fact 229.2. Mendeleev was somewhat unlucky regarding this case since he was not to know that protactinium is a member of only five "pair reversals" in the entire periodic table. This situation occurs when two elements need to be reversed, contrary to their atomic weights, in order to classify them correctly. The most clear-cut case of this effect was that of tellurium and iodine, as discussed in chapters 1 and 2.[2]

It was not until the work of Moseley in 1914 that a clear understanding of the problem was obtained. As Moseley showed, the more correct ordering principle for the elements is atomic number and not atomic weight. The justification for placing tellurium before iodine, as demanded by their chemical properties, is that tellurium has a lower atomic number.

Returning to protactinium, it appears that Mendeleev's brief predictions were broadly fulfilled since the element does indeed show an analogy with tantalum in forming Pa_2O_5 as its highest and most stable oxide. Nevertheless, protactinium also shows a strong horizontal analogy with thorium and uranium by displaying the +4 oxidation state, something that Mendeleev does not seem to have anticipated.[3] Finally, as Mendeleev correctly predicted, protactinium does indeed occur with uranium, and more specifically in pitchblende, which is the mineral from which it was eventually isolated by Lise Meitner in 1917.

Crookes

The first person to find any signs of Mendeleev's eka-tantalum, although he was not able to isolate it, was William Crookes, a well-known London-based chemist-inventor and journal editor.[4] In a paper written in the *Proceedings of the Royal Society of London* for

1899–1900, Crookes goes into meticulous detail about the experiments he has conducted on various ores of uranium. Before even examining Crookes's paper, it is worth pausing to consider the state of research into radioactivity at the turn of the twentieth century. Much of this early work was carried out in France on uranium, an element that had been known for some time.

Uranium had first been reported in 1789 by Martin Klaproth working in Berlin, while in the course of examining the ore called pitchblende. It was Klaproth who gave the element its name in recognition of the discovery of the planet Uranus by Herschel just eight years before.[5] But what Klaproth did not realize was that he had actually isolated uranium oxide, UO_2, and not the element itself. The latter task took a further fifty or so years, until Eugène Peligot, working in France in 1841, succeeded in extracting pure uranium. But more than a century had passed between Klaproth's original discovery of UO_2 and the further discovery, also in France, that uranium possesses the remarkable property of radioactivity.

Interlude on French Researchers

One of the many scientists to whom Röntgen sent his X-ray images was Henri Poincaré. Poincaré in turn showed one of these radiographs to his colleagues at the Academy of Sciences in Paris, on January 20, 1896. One of them, Henri Becquerel, a professor at the Musée d'Histoire Naturelle, took note of a remark by Poincaré on the possible link between X-rays and luminescence. On returning to his laboratory, he designed an experiment to test the hypothesis that X-ray emission and luminescence are related. In order to see if a phosphorescent body emitted X-rays, he chose a hydrated salt of uranium that he had prepared some years before. On February 20, Becquerel placed a transparent crystal of the salt on a photographic plate, wrapped between two thick sheets of black paper, and the experiment was exposed to sunlight for several hours. After development, the silhouette of the crystal appeared on the photograph, and Becquerel concluded that the phosphorescent substance emitted a penetrating radiation able to pass through black paper.

Unable to repeat such experiments in the following days because of a lack of sunshine, Becquerel put away his salt crystal, placing it by chance on an undeveloped photographic plate in a drawer. Later, he developed the plate in order to determine the amount by which the phosphorescence had decreased. To his great surprise he found that the phosphorescence had not decreased, but was more intense than it had been on the first day. Noticing a shadow on the plate made by a piece of metal he had placed between it and the salt, Becquerel realized that the salt's activity had continued in the darkness. Clearly, sunlight had been unnecessary for the emission of the penetrating rays. Could it be that just one year after the discovery of X-rays another new form of emanation was beginning to reveal itself?

Becquerel also found that the activity of his uranium salt did not diminish with time, even after several months. He also tried to use a nonphosphorescent uranium salt and found that the new effect persisted. Soon, he concluded that the emanation was due to the element uranium itself. Even after about a year had passed from when he first began his experiments, the intensity of the new rays had shown no signs of decreasing. But Becquerel was soon to move onto other scientific interests and it was left to others to explore the rays in greater detail.[6]

The Curies and Debierne

The French tradition for work on radioactivity was carried forward by Marie and Pierre Curie, who together isolated two further elements that turned out to be far more radioactive than uranium, namely polonium, followed by radium. And very soon afterward yet another radioactive element, actinium, was also discovered in France, by André Debierne (fig. 3.3).

88	89	90	91	92
Ra 1898	Ac 1899	Th 1815	Eka -Ta 1917	U 1789
Curies	Debierne	Berzelius	Meitner	Klaproth

FIGURE 3.3 Fragment table to locate these elements (atomic number, symbol, date of discovery, and discoverer).

Back to Crookes

In his paper of 1900 Crookes begins by conducting a number of experiments on various uranium compounds. He concludes by saying,

> Thus it appears that no modifications of physical or chemical condition materially affects the radio-active property of a uranium compound when, to begin with, the salt experimented on possesses it; other similar experiments show that, starting with an inactive uranium salt, nothing that can be done to it will cause it to acquire this property. It is therefore evident that, as I suspected, the radio-active property ascribed to uranium and its compounds is not an inherent property of the element, but resides in some outside body which can be separated from it.[7]

It is this search for an "outside body" that led Crookes to begin to identify a substance that he called uranium-X, which would eventually lead to the discovery of the element one place before uranium in the periodic table and which would eventually be called protactinium. It is called protactinium because on experiencing alpha decay it changes its atomic number from 91 to 89, which means that it changes from protactinium into actinium. It is, therefore, the progenitor of actinium, hence proto-actinium, which was soon abbreviated to the less awkward-sounding protactinium.

In the same article, Crookes then set about trying to see if such a body can be separated from uranium. Following a further battery of tests he established that,

> Having definitively proved that the radioactivity of uranium and its salts is not an inherent property of the element, but is due to the presence of a foreign body,* it is necessary patiently to determine the nature of the foreign body. Several radio-active bodies claimed to be new have already been extracted from pitch-blende, and experiments have been instituted to see if the newly found body UrX had similar chemical properties to those of the older active substances.

In the footnote denoted by the asterisk, Crookes gives a provisional name to the new substance by writing,

> For the sake of lucidity the new body must have a name. Until it is more tractable I will call it provisionally UrX—the unknown substance in uranium.

This was the first time that the substance that would eventually yield protactinium was given a name or label of any kind.

Crookes proceeded by examining Curie's element, polonium, to see whether it might be the same as UrX, but soon concluded that it was not. He then turned to the possibility that it might be Curie's other element, radium, and although this comparison proved more difficult, Crookes was able to conclude again that it was not. Crookes carefully examined the spectral evidence on UrX and previously obtained spectral data on polonium and radium but could not be so sure. His paper concluded somewhat poetically by emphasizing just how difficult such experiments were. The net outcome seems to be that Crookes had realized an important new substance existed but had no means of extracting it or even of categorically stating that it might be a new element rather than a previously observed one:

> Like an astronomer photographing stars too faint for his telescope to disclose, he has only to expose the plate for a sufficiently long time and the star reveals itself on development. So in the case of radioactive minerals or precipitates, if no action is apparent at the end of one hour, one may be shown after twenty-four hours. If a day's exposure will show nothing, try a week's. Considering my most active UrX does not contain sufficient of the real material to show in the spectrograph, yet is powerful enough to give a good impression on a photographic plate in five minutes, what must be its dilution in compounds, which require an hour, a day, or a week to give an action?

The Search for the Mother of Actinium and the Discovery of Brevium

The work on UX and its variations seems to have been taken up again rather enthusiastically by a number of researchers in the year 1913. It should be stressed that none of these researchers on radioactivity was setting out to deliberately discover a new element. Their goals were somewhat broader. They were engaged in clearing up the confusion of many newly discovered radioactive elements or whether or not they were really elements. The concept of isotopy was only just beginning to be clarified, due to the efforts of Soddy in particular. It had not yet been realized that atomic number rather than atomic weight was the more correct ordering principle for the elements. As a result, it was not known how many elements existed between lead and uranium. In fact, it was not even clear whether the periodic table remained valid at such high values of atomic weight. Such concerns were deepened by the fact that the rare earth elements could not be accommodated within the main body of the periodic table and needed to be listed separately.[8]

In the middle of this uncertainty, the element actinium remained mysterious, especially regarding its lineage in terms of radioactive decay. Actinium was the least understood of the recently discovered radioelements, with a still unknown atomic weight. Its chemistry was uncertain and the value of its half-life was in dispute. It was known that actinium was the start of one of the three important radioactive decay series, but it was also known that it was somehow descended from uranium. Furthermore, actinium was always found, and indeed only found, in uranium-bearing minerals and yet the link between the two elements remained mysterious.

As a result, the search began for the origin of actinium, or as it was termed in the Germanic nations, the "mother of actinium." For example, according to Lise Meitner and Otto Hahn, working in the new Kaiser Wilhelm Institute in Berlin, the problem was:

> To find that substance . . . forms the starting point for the actinium series, and to determine whether and through which intermediates actinium is derived.[9]

The situation became a little clearer following some important discoveries made in the field of radioactivity. Most importantly, some patterns in the behavior of α and β decay came to light. At about the same time, Fajans, a Polish radiochemist working in Karlsruhe, Germany, and Frederick Soddy, another radiochemist working in Glasgow, Scotland, announced the displacement laws. These were both simple and highly illuminating in efforts to sort out the profusion of new elements and decay products.

These laws stated that if an element underwent α decay it would produce another one with a charge of two units lower. On the other hand, an element undergoing β decay resulted, somewhat surprisingly, in a product with a charge of one unit higher. These new rules seemed to be correlating atomic charge, rather than weight, with an element's position in the periodic table, but matters were still uncertain.[10]

Partly on the basis of these radioactivity displacement rules, several radio-elements had been successfully placed in the periodic table. Curie's radium showed a divalency similar to that of barium and was therefore safely assigned to group II. The chemistry of thorium pointed to tetravalency and therefore meant that the element should be placed in group IV, while the hexavalency of uranium meant that it belonged in group VI. [11]

Although the placement of actinium remained doubtful, as did many of its properties, Soddy, Meitner, Hahn, and Fajans all independently concluded that the element belonged to group III. This sequence of elements that appeared to be homologous to the third row of the transition metals therefore showed a conspicuous gap representing a pentavalent element that was expected to have properties similar to tantalum in group V of the periodic table (fig. 3.4).

This is more or less where matters stood when the search for the mother of actinium began to take shape. Drawing on the new displacement rules of Fajans and Soddy, it was clear that there were two ways in principle in which actinium could form. Either it was the daughter product of radium, following β decay, or else it was formed by the β decay of the suspected new group V element, or eka-tantalum as it was dubbed by Soddy. In a paper published in *Nature* in 1913, Soddy concluded that actinium was probably formed from the suspected new element rather than from radium, since radium had never been found together with actinium (fig. 3.5).

A few months earlier Fajans and Göhring in Karlsruhe, Germany, had discovered UX_2 a daughter product of the β emission of UX_1, which had been identified as an isotope of thorium. A quick glance at fig. 3.1, together with the displacement laws, shows that UX_2 was none other than the suspected new element or at least one isotope of it. If one were taking a narrow view of the discovery of element 91,

Groups II	*III*	*IV*	*V*	*VI*
Ra	Ac	Th	?	U

FIGURE 3.4 Fragment table showing original placement of elements radium to uranium.

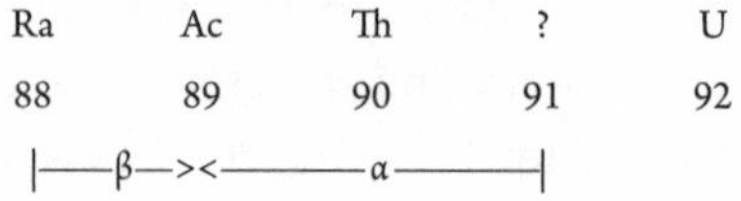

Ra	Ac	Th	?	U
88	89	90	91	92

|—β—><————α————|

FIGURE 3.5 Two paths to the formation of Ac, from radium by β decay or from the suspected new element by α decay. Atomic numbers have been used to identify elements whereas the concept had yet to be solidified in 1913.

one would have to identify it with this discovery by Fajans and Göhring.

Although Crookes may have worked with UX, he clearly did not identify a new element. Fajans and Göhring, however, with the aid of the displacement laws, could quite confidently claim this as their prize. Moreover, they confirmed their suspicion by carrying out some chemical tests, finding that the suspected new element could be separated from its parent thorium mineral by using tantalic acid.

They named their new element brevium for the very short half-life of 1.17 minutes that it possessed. But the brevity of this half-life also contributed to the fact that this element name has not survived. There was a convention among radiochemists that the name of an element should be given to the discoverers of a stable isotope rather than an unstable one. As it turned out, a far, far more stable isotope of the new element was discovered a few years later.

But let us get back to the search for the mother of actinium. According to Fajans and Göhring, their own brevium was a β emitter that formed an isotope of uranium (^{234}U). This meant that it could simply not be the mother of actinium. Although element 91, in a very short-lived form, had technically been discovered, the more pressing problem of the origin of actinium had not, although the possibilities had been narrowed down.

At this point, Meitner and Hahn in Berlin proposed another plausible mechanism for the formation of actinium from the suspected new element. Here is how Soddy explained Meitner and Hahn's view at the end of his 1913 paper in *Nature*:

> So far the experiments appear to disprove the possibility that actinium can be formed from radium. Similar arguments to those above may be used to show that it cannot be a primary radio-element, and so its origin remains still a mystery.

However, Soddy adds a final note:

> In the current number of *Physikalische Zeitschrift* (p. 752) Hahn and Meitner modify my original suggestion and suppose that the branching of the uranium series takes place at uranium-X, two simultaneous β-ray changes occurring, which produce two eka-tantalums, one the known short-lived β-ray giving product [brevium] and the other a still unknown long-lived α-ray-giving parent of actinium. It is almost the only other alternative remaining to be tested, and it should not be difficult to settle by experiment.

As it turned out, this is precisely how Hahn and Meitner, or rather Meitner working almost completely alone, succeeded in discovering a long-lived isotope of the suspected new group V element.

Meitner and Hahn's Path to Protactinium

Although the discovery of a long-lived isotope of the suspected element in group V, or eka-tantalum, is usually attributed to Lise Meitner (fig. 3.6) and Otto Hahn, it is quite clear from the lengthy correspondence between these two collaborators that the vast majority of the work was carried out by Meitner. This case is made especially interesting due to the difficult background conditions Meitner experienced while carrying out this work during World War I. The research is described in detail in the correspondence in a way

FIGURE 3.6 Austrian postage stamp depicting Lise Meitner. Courtesy of Professor Daniel Rabinovich of Department of Chemistry, University of North Carolina, Charlotte, NC.

that one does not see in published research papers associated with most scientific discoveries.

Although Meitner and Hahn had recently moved into the Kaiser Wilhelm Institute for Chemistry in Berlin, Otto Hahn soon went off to fight in the war and Meitner was left to carry out all experiments on her own. Not only did she lose her long-term companion and collaborator but, as Meitner frequently mentions in the correspondence, she also lost all students and technical assistants.

Given the weakness of the radioactivity associated with the precursor of actinium, it was fortunate that Meitner and Hahn had recently moved into a new laboratory, since their former lab had become contaminated by five or so years of work on radioactive isotopes. They were well aware of this problem and Meitner, in particular, went to great lengths to instruct all students and technicians in their new laboratory of the need for strict procedures to avoid contamination.

Meitner and Hahn attacked the problem by means of two approaches. For the first approach, they attempted to work with uranium salts, which had been extracted from their ores some twenty-five years previously. The second tactic involved the cruder pitchblende, the ore that contained uranium and several other radioelements. In both cases, the sought-for slow α decay was masked by more rapid decays due to thorium and polonium. It was clear to Meitner and Hahn that they would need to monitor the samples for several years in the hope that the rapid extraneous decays would decrease to reveal the much slower decay due to the mother substance of actinium. The prepared samples were therefore mounted under several electroscopes and left for periods of several years.

In 1914, Meitner and Hahn had a lucky break, in discovering that the use of nitric acid as a solvent for pitchblende formed a residue of silicon dioxide that contained only very small amounts of the usual contaminants of polonium, thorium, and bismuth. By contrast, the residue contained elements in group V and so presumably the missing new element, namely the mother substance of actinium. Meitner and Hahn did not share this method with their competitors since they were aware that it would confer a big advantage to anyone who possessed knowledge of it.

At this point in 1914, the war broke out. Otto Hahn was conscripted immediately and served in a notorious poison gas unit headed by Fritz Haber. Meanwhile, Meitner volunteered her services as an X-ray nurse. After a brief spell, however, she grew weary of the work and returned to doing science at the institute in Berlin. Meitner returned to find that Haber's people had helped themselves to a great deal of equipment from many of the labs but, fortunately, none of the electroscopes from her experiments searching for eka-tantalum. In letters to Hahn she wrote:

> The Haber people treat us, of course, like captured territory—they don't take what they *need* but what they *like*. Who will guarantee that they won't come over here, and then everything will be lost . . . I

> shall do everything to prevent it, we have had measurements running for such a long time.[12]

By January of 1917, when Hahn returned to Berlin for a brief spell, they measured the radioactivity, which seemed to indicate a definite amount of actinium precursor. After Hahn went back to the army, Meitner was left to continue alone again, while badly needing more pitchblende ore to pursue the radioactivity measurements. In a letter of February 1917 Meitner describes some of the difficulties she is enduring:

> I have ordered the vessels for our actinium experiments, will get them in a few days and will begin right away... be well and please don't be angry about the delays with the pitchblende. Believe me, it is not because of a lack of will, but because of lack of time. I can't very well do as much work as the three of us did together. Yesterday I bought 3 meters of rubber tubing for 22M!! I got quite a shock when I saw the bill.[13]

A short time afterward she was able to report to Hahn that she had proof that the silica residue in the pitchblende contained the new substance and that pitchblende was indeed a good source from which to extract the new element. But Meitner was starting to need additional amounts of the pitchblende. She traveled to Vienna in neighboring Austria in order to request this material from the head of the Radium Institute and a longtime colleague, Stephan Meyer. Although Meyer provided her with a kilogram of pitchblende from which the uranium had been extracted, she also faced issues with the authorities because an export ban had been imposed on all goods moving from Austria to Germany. Once more she appealed to Meyer for the necessary clearance, which was eventually granted to her due to his intervention. In the meantime, Meitner also approached Friedrich Giesel, a German producer of radium who was willing to oblige with some more pitchblende.

In another letter to Hahn, Meitner informs him that Fajans has just published a paper defending his claim for brevium and denying any claims by Soddy to having discovered eka-tantalum. Meanwhile, Meitner's own experiments were progressing as scheduled, which she conveyed to Hahn by writing:

> The larger samples from Giesel's pitchblende also showed increasing alpha activity, indicating the accumulation of actinium emanation.[14]

Each successive experiment was providing Meitner with increasing confirmation that she was succeeding in finding the mother substance of actinium. Again she turned to Giesel for more raw material as it became necessary to measure the range of the observed α activity and the rate at which the active deposits were being generated. This time she traveled to Braunschweig to speak to Giesel personally and was rewarded by receiving more than a hundred pounds of the pitchblende residues with which to continue her work.

The final stages were duly completed, with a little assistance from Hahn while on one of his occasional visits to Berlin. In March of 1918 Meitner and Hahn submitted an article entitled "The Mother Substance of Actinium, a New Radioactive Element of Long Half-Life." This included the simple claim:

> We have succeeded in discovering a new radioactive element, and demonstarating that it is the mother substance of actinium. We propose, therefore, the name protoactinium.[15]

As an interesting aside, Meitner was able to inform Meyer of her triumph after the paper had been published and received from him the following friendly message:

> In your letter you pose a terribly difficult question about protoactinium. I would prefer the names Lisonium, Lisottonium, etc. and

> I therefore propose the symbol Lo, but unfortunately these are unsuitable if one wants general acceptance... Although I still prefer Lisotto, it is much more significant to have discovered Pa or Pn than to come up with the most beautiful name.[16]

Meitner and Hahn also had to face a potentially delicate situation with Fajans, who, as Meitner was well aware, was claiming priority for another isotope of the element that he had observed some five years earlier. Rather surprisingly to all concerned, however, Fajans seems to have readily given up his claim in the face of Meitner and Hahn's article of 1918. This was because of the common practice, mentioned earlier, of denoting an element by its longest lived isotope. Whereas Fajan's brevium had a half-life of just 1.7 minutes, the Meitner-Hahn isotope had a relatively gargantuan half-life of 32,500 years! Clearly, there was no contest and Fajans felt compelled to give way.

Then there was a third set of contenders, the Scottish duo of Soddy and his young colleague Cranston, who at the time only had a bachelor's degree. In fact, Soddy and Cranston had actually beaten the Germans in terms of publication since their paper was submitted in December of 1917, while the German paper was sent to *Physikalische Zeitschrift* in March of 1918. Whereas Fajans had observed brevium, which is the very short-lived isotope ^{234}Pa, Soddy and Cranston as well as Meitner and Hahn had observed the exceedingly long-lived isotope of ^{238}Pa. But the team from Scotland had only formed a very small amount of the isotope and, more crucially, had not been able to characterize its properties to anywhere the same extent as Meitner had. To their credit, Soddy and Cranston readily conceded priority to Meitner and Hahn.[17]

Lise Meitner's and Otto Hahn's two most famous discoveries were those of protactinium and later nuclear fission, a process that would change the world and especially the balance of military power. Interestingly, both discoveries occurred while one of them was away from the laboratory where they worked during successive world wars.

In the case of protactinium, discovered during World War I, most of the experiments were conducted by Meitner while Otto Hahn was away from Berlin and engaged in research into poison gases. In the case of nuclear fission, Meitner, a Jew, had fled from Berlin to Sweden, from where she maintained a regular and frequent correspondence with Hahn. Experiments with uranium revealed that a completely unrelated element of barium was being formed.

At first Meitner directed Hahn to the periodic table, telling him that something had to be wrong, given that uranium was number 92 and barium was much smaller, at number 56. As these observations persisted, Meitner, along with her nephew Otto Frisch, finally realized what was happening. Unlike the previously observed radioactive decay processes that produced changes in atomic numbers of one or two units, here was a completely new process that led to the breakup of a large nucleus and the formation of substantially smaller nuclei such as barium and krypton.

But although the discovery of nuclear fission was essentially a German affair, the possibility of using this process to build a nuclear weapon was conveyed to the United States by Bohr and later Einstein. This in turn led to the establishment of the Manhattan Project and the eventual deployment of the two atomic bombs over Japan that prompted a swift end to World War II.

Curiously, another major character who features in this book's "tale of seven elements," Ida Noddack, the codiscoverer of rhenium, had anticipated the process of nuclear fission but had been ignored, some believe due to her previous erroneous claim for also having discovered element 43, another of our "seven elements."

Between the two wars, Meitner and Hahn did not collaborate, and the physicist, Meitner, became a good deal more prominent than the chemist, Hahn. In 1934, Meitner became fascinated by reports that Enrico Fermi had created some new elements beyond uranium. Meitner appealed to her old friend Hahn because she needed a radiochemist able to analyze the fragments obtained in such nuclear experiments. After Hahn accepted, they worked together to bombard nuclei of uranium with neutrons in the same way that Fermi had, and in 1937 a third member, Fritz Strassmann, was added to the team. Throughout this collaboration Meitner was regarded as the intellectual leader of the team.

While the Berlin team, as well as Fermi in Rome and the Joliot-Curies in Paris, were all conducting experiments that actually involved nuclear fission, none of them realized this fact. They all thought that they were forming artificial elements that were heavier than uranium and this ignorance persisted for a full four years. The fact that the heavy uranium nucleus was in fact splitting into two considerably lighter nuclei still eluded them.

When Strassmann reported what he thought were barium atoms in 1936, an incredulous Meitner asked him to check his chemistry. After reports from Paris indicated the possible formation of lanthanum, another middle-sized nucleus, it was Hahn who poured cold water over the claim.

It was at this point that Ida Noddack suggested that the uranium nucleus was splitting into two medium-sized nuclei, but failed to perform any supporting experiments.

By this time, the German anti-Jewish campaign was gaining strength. When Hitler first came to power and began to impose his anti-Jewish policies, Meitner, being of Austrian rather than German nationality, was initially unaffected. This was soon to change. Under the newer policies, regardless of whether or not she was a German citizen, Lise Meitner was prevented from teaching and even from attending any research seminars.[18]

In 1935, Niels Bohr arranged for Meitner to spend a year in his institute in Copenhagen but she turned it down, believing that she could remain in Germany. In 1938, Germany invaded Austria in the buildup to the war, with the result that Meitner was now considered a German Jew and therefore fully subject to anti-Jewish regulations. Soon, Hahn, who was publicly opposed to Nazi policies, bowed to pressure and asked Meitner to resign from the institute in Berlin.

Matters then moved very quickly. Meitner learned that Jews would not be allowed to leave Germany and that she would have to act quickly in order to move to a neutral country. The Dutch physicist and codiscoverer of hafnium (chapter 4) Dirk Coster, traveled to Berlin in order to personally escort Meitner to neighboring Holland. Meitner left Berlin on July 13, 1938, after packing just a few personal belongings in order not to arouse any suspicion. After a few weeks in Holland and another short spell at Bohr's institute in Copenhagen, she moved again to accept an offer from a physics research institute in Stockholm, Sweden. This was a decision that Meitner would come to regret because she had little equipment for experiments and found herself without any like-minded colleagues. Nevertheless, she maintained a regular correspondence with Hahn, who had stayed in Berlin.

Returning to scientific matters, Hahn and Strassmann's further analysis of the decay products of uranium led to the view that radium (element 88) had been produced. On hearing this report Meitner asked her German colleagues to repeat their experiments and to check the decay products more carefully. It was these further experiments that culminated on December 19, 1938, with Hahn and Strassmann's realization that they had, in fact, produced the much smaller nucleus of barium. Still somewhat confused about their results, however, Hahn wrote to Meitner saying:

> There is something about the "radium isotopes" that is so remarkable that for now we are telling only you... Our Ra isotopes act like Ba.

At first Meitner was equally puzzled but responded with:

> One cannot say without further consideration that it (Ba) is impossible.

"On receiving this encouragement, Hahn and Strassmann immediately sent an article off to a journal."[19]

> As chemists, we must rename [our] scheme and insert the symbols Ba, La, Ce in place of Ra, Ac, Th. As nuclear chemists closely associated

> with physics, we cannot yet convince ourselves to make this leap, which contradicts all previous experience in nuclear physics.

Meanwhile, back in Sweden on December 30, Meitner met up with her nephew Otto Frisch, also a physicist, and decided to go on a cross-country skiing trip. During this excursion the two of them discussed how a uranium nucleus might possibly be producing a nucleus as light as that of barium. According to the model of the nucleus that was current at the time, the nucleus behaved like a liquid drop. Meitner and Frisch realized that such a nuclear drop could become distorted into an oval shape and consequently might separate into two smaller "drops," hence the explanation for the formation of the smaller nucleus of barium. They also calculated that such a fission process, as Frisch proposed to call it, could release twenty million times the energy of an equivalent amount of TNT.

As science historian Sharon McGrayne notes:[20]

> Sixty years old and officially retired, Meitner had explained one of the greatest discoveries of the century.

But Why No Nobel Prize for Meitner?

As the quotes above show, Hahn and Strassmann would have been lost without Meitner's correct interpretation that nuclear fission was taking place in their experiments. Nevertheless, it was Hahn alone who received the Nobel Prize for this discovery and, to add insult to injury, he never acknowledged Meitner's pivotal role. Hahn even stated that chemistry rather than physics had been responsible for understanding nuclear fission (fig. 3.8).

Bohr learned of the discovery the day after it took place in Sweden from Otto Frisch after the latter's return to the institute in Copenhagen where he held a position in physics. A few days later, Bohr departed for a visit to the United States, where he told a number of scientists the news. Although he was keen to support Meitner and Frisch's role in the discovery, Bohr was not able to do so very successfully, partly because it took some weeks before the two discoverers managed to publish a paper on the subject, by which time Hahn's article had been out for several weeks. Less than a month had passed before Hahn began to claim that physics had actually hampered the discovery of nuclear fission and that it had been a triumph for chemistry alone, in an obvious slur to Meitner's contribution.

But by what is perhaps a nice touch of irony, an element that had been provisionally called hahnium was eventually renamed as meitnerium, albeit several years after Meitner's death in 1968. Hahn had effectively been deprived of having an element named after him. Too little too late perhaps, but quite significantly the suggestion came from a German, Peter Armbruster, the leader of the team that had synthesized the element.[21]

Chemistry of Protactinium

Protactinium is a silvery-white metal that slowly loses its luster on exposure to the atmosphere. Meitner and Hahn were the first to explore the chemistry of the element and found it to be similar to tantalum, which is the element above it in early periodic tables.[22] In the current periodic table, protactinium is, however, not placed below tantalum, since it is now regarded as an actinide rather than a transition element. The change came about during the 1940s when Glenn Seaborg suggested that although the first few elements, in what was the fourth transition series, seemed to behave like their upper homologues, they really belonged to a separate series of inner transition elements.

In 1945, Seaborg suggested that the elements beginning with actinium, number 89, should be considered as a rare earth series, whereas it had previously been supposed that the new series of rare earths would begin after element number 92, or uranium (fig. 3.7).

Seaborg's new periodic table revealed an analogy between europium (63) and gadolinium (64) and the as yet undiscovered elements 95 and 96, respectively. On the basis of these analogies, Seaborg succeeded in synthesizing and identifying the two new elements, which were subsequently named americium and curium. A number of further transuranium elements have subsequently been synthesized (see chapter 10).[23]

The standard form of the periodic table has also undergone some minor changes regarding the elements that mark the beginning of the third and fourth rows of the transition elements. Whereas older periodic tables show these elements to be lanthanum (57) and actinium (89), more recent experimental evidence and analysis have put lutetium (71) and lawrencium (103) in their former places.[24] It is also interesting to note that some even older periodic tables based on macroscopic properties had anticipated these changes.

One of the properties that protactinium does share with tantalum is the formation of an oxide with the same stoichiometry, Pa_2O_5 like Ta_2O_5. In addition to reacting with oxygen, protactinium reacts with steam and acids but not alkalis. It is also a superconductor at temperatures below 1.4 K and occurs in two of the major radioactive decay series. While ^{231}Pa is produced naturally by the decay of ^{234}U, ^{234}Pa is the decay product of ^{238}U. In addition, there are a further 19 known isotopes of the element, all of which have half-lives of less than one month.

The first person to prepare Pa_2O_5 was the German nuclear chemist Aristid von Grosse, who did so in 1927. Seven years later he was the first to extract protactinium in elemental form, starting from another compound, protactinium iodide. Grosse was also ahead of his time in questioning the analogy between protactinium and tantalum that had been generally accepted since the time of Mendeleev. Writing in 1930, Grosse says:[25]

> Until recently all attempts to concentrate and isolate the new element [protactinium] have been unsuccessful. These attempts have

																H	He
Li	Be											B	C	N	O	F	Ne
Na	Mg											Al	Si	P	S	Cl	Ar
K	Ca	Sc	Ti	V	Cr	Mn	Fe	Co	Ni	Cu	Zn	Ga	Ge	As	Se	Br	Kr
Rb	Sr	Y	Zr	Nb	Mo	Tc	Ru	Rh	Pd	Ag	Cd	In	Sn	Sb	Te	I	Xe
Cs	Ba	RE	Hf	Ta	W	Re	Os	Ir	Pt	Au	Hg	Tl	Pb	Bi	Po	At	Rn
Fr	Ra	Ac	Th	Pa	U												

rare earths	La	Ce	Pr	Nd	Pm	Sm	Eu	Gd	Tb	Dy	Ho	Er	Tm	Yb	Lu

																H	He
Li	Be											B	C	N	O	F	Ne
Na	Mg											Al	Si	P	S	Cl	Ar
K	Ca	Sc	Ti	V	Cr	Mn	Fe	Co	Ni	Cu	Zn	Ga	Ge	As	Se	Br	Kr
Rb	Sr	Y	Zr	Nb	Mo	Tc	Ru	Rh	Pd	Ag	Cd	In	Sn	Sb	Te	I	Xe
Cs	Ba	LA	Hf	Ta	W	Re	Os	Ir	Pt	Au	Hg	Tl	Pb	Bi	Po	At	Rn
Fr	Ra	Ac															

LA	La	Ce	Pr	Nd	Pm	Sm	Eu	Gd	Tb	Dy	Ho	Er	Tm	Yb	Lu
AC	Ac	Th	Pa	U	Np	Pu									

FIGURE 3.7 Medium-long periodic table before and after Seaborg's modification.

been based on the assumption that an analogy exists between the properties of ekatantalum and tantalum similar to that between radium and barium. This assumption has been supported by the authority of many prominent chemists and was one of the principal reasons for failure in attempts to isolate ekatantulum. In November 1926 the author presented Professor O. Hahn with a report entitled "The Properties of Protoactinium and Its Compounds Calculated According to the Periodic Law." The principal deductions were as follows:

Element 91, ekatantalum, will have its own characteristic properties and analytical reactions, differing from those of tantalum and columbium, just as its neighbours, thorium and uranium differ greatly in their properties from those of their lower homologues hafnium and zirconium and tungsten and molybdenum.

Between 1959 and 1961, the Atomic Energy Authority in the United Kingdom succeeded in extracting 125g of protactinium, which remains the world's largest stockpile of the element. This feat was accomplished by starting with 60 tons of waste uranium minerals

FIGURE 3.8 Otto Hahn, codiscoverer of protactinium and also nuclear fission. Courtesy of Professor Daniel Rabinovich of Department of Chemistry, University of North Carolina, Charlotte, NC.

that were purchased for a sum of about $500,000. Meanwhile, the resulting protactinium was sold at a price of almost $3,000 per gram.[26]

Applications

Protactinium is one of a very few elements among the first 92 in the periodic table that has virtually no applications. This is because, as stated throughout the chapter, the element is extremely rare, not to mention highly toxic and highly radioactive. Nevertheless, one particular isotope, ^{231}Pa, has found a very specific use in scientific research. It has been utilized to study the movement of ocean waters in geology and a field called paleoceanography, or the study of ancient oceans.

Ocean sediments as old as 175,000 years have been dated more accurately by using ^{231}Pa than they could be dated by more conventional dating techniques. In addition, the movement of the oceans, as they were melting following the last Ice Age, have been studied through the use of the same isotope. The method exploits the fact that compounds of thorium and protactinium, elements 90 and 91, respectively, show different rates of sedimentation as they precipitate out of the water in the oceans. While each of ^{230}Th and ^{231}Pa are individually good indicators of the age of a sediment, the ratio of the two quantities presents a better measure because considering this ratio avoids any problems that might occur if either of the rates of decay occur in a nonuniform manner.

Chapter 4

Element 72—Hafnium

H																	He
Li	Be											B	C	N	O	F	Ne
Na	Mg											Al	Si	P	S	Cl	Ar
K	Ca	Sc	Ti	V	Cr	Mn	Fe	Co	Ni	Cu	Zn	Ga	Ge	As	Se	Br	Kr
Rb	Sr	Y	Zr	Nb	Mo	Tc	Ru	Rh	Pd	Ag	Cd	In	Sn	Sb	Te	I	Xe
Cs	Ba	Lu	**Hf**	Ta	W	Re	Os	Ir	Pt	Au	Hg	Tl	Pb	Bi	Po	At	Rn
Fr	Ra	Lr	Rf	Db	Sg	Bh	Hs	Mt	Ds	Rg	Cn		Fl		Lv		

La	Ce	Pr	Nd	Pm	Sm	Eu	Gd	Tb	Dy	Ho	Er	Tm	Yb
Ac	Th	Pa	U	Np	Pu	Am	Cm	Bk	Cf	Es	Fm	Md	No

FIGURE 4.1 Showing the position of element 72, eventually named hafnium in the periodic table.

The story concerning the discovery and isolation of element 72 bears all the characteristics of controversy and nationalistic overtones that seems to characterize many of our seven elements. On one hand, it seems odd that there should be so much controversy associated with these elements given that Moseley's method had apparently provided an unequivocal means through which elements could be identified as well as a way of knowing just how many elements remained to be discovered.

On the other hand, perhaps it was precisely because the problem of the missing elements became so clearly focused on a few elements, with known atomic numbers, that the stakes became higher than they would have been if the number of elements remaining to be discovered had been uncertain, as they were in pre-Moseley times.

Element 72 (fig. 4.1) was clearly anticipated, although not as such, even in Mendeleev's earliest table of 1869. As fig. 4.2 shows, Mendeleev considered that an as yet undiscovered element with an atomic number of 180 should be a homologue of zirconium (The modern accepted value is 178.50). This fact may not seem very significant and yet we will see, as the story of this chapter unfolds, that it amounts to Mendeleev predicting that this

преимущественно найдти общую систему элементовъ. Вотъ этотъ опытъ:

			Ti=50	Zr=90	?=180.
			V=51	Nb=94	Ta=182.
			Cr=52	Mo=96	W=186.
			Mn=55	Rh=104,4	Pt=197,4
			Fe=56	Ru=104,4	Ir=198.
			Ni=Co=59	Pd=106,6	Os=199.
H=1			Cu=63,4	Ag=108	Hg=200.
	Be=9,4	Mg=24	Zn=65,2	Cd=112	
	B=11	Al=27,4	?=68	Ur=116	Au=197?
	C=12	Si=28	?=70	Sn=118	
	N=14	P=31	As=75	Sb=122	Bi=210
	O=16	S=32	Se=79,4	Te=128?	
	F=19	Cl=35,5	Br=80	I=127	
Li=7	Na=23	K=39	Rb=85,4	Cs=133	Tl=204
		Ca=40	Sr=87,6	Ba=137	Pb=207.
		?=45	Ce=92		
		?Er=56	La=94		
		?Yt=60	Di=95		
		?In=75,6	Th=118?		

FIGURE 4.2 Mendeleev's first periodic table of 1869. D. I. Mendeleev, Sootnoshenie svoistv s atomnym vesom elementov, *Zhurnal Russkeo Fiziko-Khimicheskoe Obshchestv*, 1, 60–77, 1869, table on p. 70.

element would be a transition metal rather than a rare earth. But Mendeleev was not really in a position to make such a statement since the nature and number of rare earth elements was unknown in his day. Indeed, the problem of the rare earths was one of the most acute challenges to his periodic system and one that he personally never resolved.[1]

Sometime later, Julius Thomsen, a chemistry professor at the University of Copenhagen and incidentally the chemistry instructor to the physicist Niels Bohr, published a periodic table in which he too included a missing element that was a homologue of zirconium (fig. 4.3).

Suffice it to say that there was a general consensus among chemists that on the basis of the periodic table there should exist an element before tantalum that would be a homologue of zirconium. The trouble was that a further two elements, now numbered 70 and 71, respectively, were also missing at this time.

Elements 70 and 71 were first isolated in 1907 by two independent researchers. The first was Georges Urbain, one of the leading French chemists of the era who specialized in experimental work on the rare earth elements. By analyzing the element ytterbium, which had been previously identified by Marignac, Urbain claimed to find two elements instead of one, and called them neoytterbium (new ytterbium) and lutecium, respectively.[2]

The second claim to the discovery of elements 70 and 71 came from the Austrian researcher Carl Auer von Welsbach, who suggested calling the elements aldebaranium and cassiopium. Urbain was eventually credited with the discovery of the elements and won the right to name them, although the names were later changed so that neoytterbium went back to ytterbium and the spelling of lutecium was changed to lutetium.[3]

The connection to element 72 began to develop after this point because Urbain then went on to suspect that there might even be a third element lurking in Marignac's original one element. In 1911, Urbain announced the discovery of what he

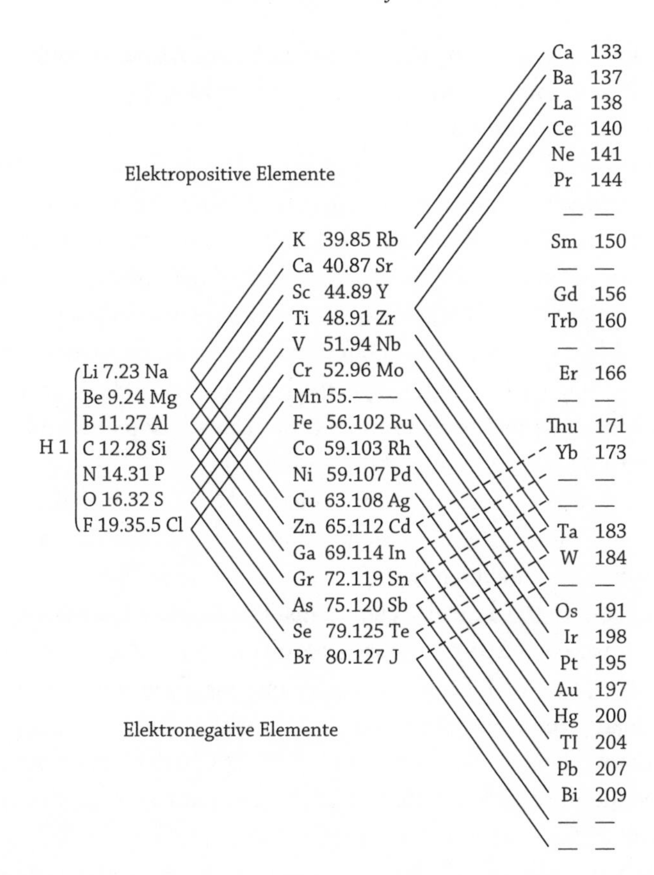

FIGURE 4.3 Thomsen's pyramidal periodic system. J. Thomsen, Systematische Gruppierung der chemischen Elemente. *Zeits für Anorgische Chemie,* 1895, 9, 190–193.

believed was this third rare earth, namely element 72, choosing to call it celtium.[4]

Soon afterward, in England, Henry Moseley developed his X-ray method for the identification of elements and for predicting precisely which elements remained to be discovered (chapter 2). In order to solidify his claim to having discovered a third element, Urbain made the trip from Paris to Oxford in order to allow Moseley to carry out his definitive test. After only a couple of days, Moseley concluded that Urbain's sample was not, in fact, the missing element in question and Urbain capitulated.

Everything was then quiet regarding element 72 for something like eleven years, until Urbain teamed up with the X-ray spectroscopist Dauvillier.[5] Dauvillier reexamined Urbain's 1911 sample and claimed to have discovered two faint X-ray lines lying almost exactly where they should have been on the basis of Moseley's law. The problem was that others could not even see these two lines. For example, Siegbahn, a leading spectroscopist who had further developed Moseley's methods, examined the Dauvillier plates and concluded that no lines were actually present.

Meanwhile, theoretical physicists had been approaching the problem from a quite different direction. On the basis of his atomic theory, Bohr had concluded that element 72 should not be a rare earth element but in fact a quadrivalent transition element. His reasons for doing so were not entirely theoretical and he admitted as much, along with citing the work of his chemistry mentor Thomsen as grounds for thinking the element should be a transition metal analogous to zirconium. He also cited the British chemist Charles Bury, who had independently concluded that element 72 was not a rare earth but an element analogous to the transition metal zirconium.[6]

Two researchers—Dutchman Coster and Hungarian Hevesy (fig. 4.4)—working at Bohr's institute in Copenhagen decided to try to settle this question in an experimental manner. On the basis of the older chemical predictions and Bohr's more recent predictions, they argued that if the new element were present it should perhaps occur along with zirconium. They obtained some Norwegian zirconium ores and within a few weeks had succeeded in observing not just two, but six, X-ray lines that were in far better agreement than Urbain's with the frequencies expected on the basis of Moseley's law. They proposed the name of hafnium from the Latin name for Copenhagen (hafnia), the city in which the element had first been detected.[7]

FIGURE 4.4 George Hevesey, one of the discoverers of hafnium. Courtesy of Professor Daniel Rabinovich of Department of Chemistry, University of North Carolina, Charlotte, NC.

Coster and Hevesy then went into print in a paper in which they also criticized the claims of Dauvillier and Urbain. Among other things, they pointed out that the two French lines were not close enough to the expected frequencies. More fully, they wrote,

> In the *Comptes Rendus* of the Paris Academy for May 22, 1922, Dauvillier announced the detection by means of X-ray spectroscopy of the element 72 in a mixture of rare-earth metals. This element was identified by Urbain with a rare-earth element, which he called celtium, the presence of which he had previously suspected in the same sample. For different reasons, however, we think that Dauvilllier's and Urbain's conclusions are not justified. It appears from Dauvillier's paper that at any rate the quantity of the element 72 in the sample, if present, must have been so small that it seems

> very improbable that the element 72 should be identical with the element which in former papers Urbain claims to have detected in the same sample by the investigation of the optical spectrum and of the magnetic properties. The only lines which Dauvillier claims to have detected are the lines $L_{\alpha 1}$ and $L_{\beta 2}$, both of which he finds to be extremely faint (extrêmement faible). The wavelengths he gives, however, for these lines are about 4 X.u. (1 X.u. = 10^{-11} cm.) smaller than those which are obtained by rational interpolation in the wavelength tables... for the elements in the neighbourhood of 72.[8]

They followed this by casting further doubt on the French claim.

> From a theoretical point of view it appears very doubtful the element 72 should be a rare-earth. It was announced in 1895 by Julius Thomsen from Copenhagen that from the general consideration of the laws of the periodic system we must expect between tantalum, which in many compounds possesses 5 valencies, and the tri-valent rare-earths, a tetra-valent element homologous to zirconium. The same view has also recently been put forward by Bury on the basis of chemical considerations, and by Bohr on the basis of his theory of atomic structure. It is one of the most striking results of the latter theory, that a rational interpretation of the appearance of the rare-earth metals in the periodic system could be given. For these elements, according to Bohr, we witness the gradual development of the 4-quantum electrons from a group containing 18 electrons into a group of 32 electrons, the number of electrons in the groups of 5- and 6-quantum electrons remaining unchanged. Bohr was able to conclude that in the element lutecium (71) the group of 4-quantum electrons is complete, and we consequently must expect that in the neutral atom of the next element (72) the number of electrons moving in 5- and 6-quantum orbits must exceed that in the rare-earths by one. The element 72 can therefore not be a rare-earth but must be an homologue of zirconium.[9]

The outcome of this paper was that it sparked off one of the most bitter and acrimonious priority disputes in twentieth-century science. On one side of the debate were the French scientists that of course included Dauvillier and Urbain but also others such as Maurice de Broglie.[10] In addition, a number of British chemists defended the French claim and persisted in calling the element celtium for quite a period of time. The only significant supporter of the Danish claim was Ernest Rutherford, who was actually a New Zealander and who had been a mentor for Bohr while the Dane had spent a postdoctoral year in England.

The cause of the partisanship in other quarters is not difficult to understand. The early 1920s were the years immediately following the Great War, and the victors, France and Britain, still resented German scientists and continued to prevent them from attending scientific meetings. In addition, the debate was taking place at a time when the French–Belgian alliance had occupied the Ruhr district of Germany. The Danes were, of course, not Germans, but were regarded as guilty by association, both through geographical proximity and the fact that Denmark had remained neutral during the war. Ironically, neither discoverer of hafnium—Coster nor Hevesy—was German or even Danish. Nevertheless, they were treated as the enemy because of the location of the institute in which the discovery was made.[11]

Upon receiving an article on hafnium submitted to the British journal *Chemical News*, the then editor W. P. Wynne made the following astonishing response:

> We adhere to the original word celtium given to it by Urbain as a representative of the great French nation which was loyal to us throughout the war. We do not accept the name which was given it by the Danes who only pocketed the spoils of war.[12]

The first published reaction to the Coster and Hevesy article announcing the discovery of hafnium came unexpectedly from

London rather than from Urbain and Dauvillier in Paris. Alexander Scott, a chemist and mineralogist from the British Museum in London, had analyzed some samples of black sand from New Zealand between the years of 1913 and 1915 and had believed that he had discovered a new element. Although he had not published his findings, he was now claiming to have discovered element 72 before Coster and Hevesy. Scott proposed to call his element oceanium, after oceania, the region from which the mineral sample originated. The further basis for his claim was that he believed to have obtained the atomic weight of the element as 144 in 1918.[13] Scott was thus at least supporting Coster and Hevesy in the notion that the element belonged in the same group as titanium and zirconium.

The Englishman was encouraged to send his "new element" to Copenhagen for analysis, which he readily agreed to. He then wrote to Coster and Hevesy in somewhat grandiose terms:

> the whole scientific world will wait your examination with breathless interest.[14]

Unfortunately for Scott, the men from Copenhagen could not find any sign of a new element. Somewhat out of politeness to Scott, they nevertheless offered to continue working on the black sand in order to search for other possible new elements such as the still missing elements 61 and 75. It was several years before Scott finally withdrew his claim. But even then the British press continued to argue the case for oceanium, including a rather patriotic editorial in *The Times* of London.

> Science is and doubtless should be, international, but it is gratifying that the chemical achievement, the most important since the late Sir Wm. Ramsay isolated helium in 1895, should have been the work of a British chemist in a London laboratory. The element 72 which was thought to be, if existing, exceedingly rare and the properties of which were calculated by Danish chemists was actually discovered by Alexander Scott.[15]

A somewhat embarrassed Rutherford, in Cambridge, felt compelled to write to Bohr, saying,

> we need pay no attention to such irresponsible utterances: things are getting quite lively over the new element...I will see you and your people get a square deal.

A more serious challenge to the Copenhagers came, a little later, from Urbain and Dauvillier, who certainly did not let the matter rest with Hevesy and Coster's rebuttal in *Nature* magazine. They responded, rather robustly, that Coster and Hevesy's work should be regarded merely as a detection of their own rare earth element, celtium, in zirconium minerals but definitely not as a discovery of a new element.

They granted that the work of Coster and Hevesy had produced

> des resultats très important

but added

> il est seulement regrettable que MM. Coster et Hevesy se soient efforces de jeter le discrédit sur nos propres résultats.

Coster and Hevesy responded to this charge of plagiarism by pointing out that hafnium could not be the same as Urbain's celtium. They explained that their own sample's x-ray lines suggested an element content of about 0.01 percent. Meanwhile, the elemental content of Urbain's samples would have had to have been considerably higher since Urbain had claimed that he could detect a gradual change in magnetic susceptibility. In addition, Coster and Hevesy pointed out that subsequent analyses of the chemical properties of their hafnium had shown similarities with the transition element zirconium and not to rare earth elements as Urbain had assumed celtium to be. Finally, Coster and Hevesy's colleagues in Copenhagen had observed the optical spectrum of hafnium and found it markedly different from the one that Urbain had originally reported for his celtium.[16]

Confronted with this evidence, which was bolstered by further examination of hafnium's optical spectrum, Urbain was gradually forced to admit that his original claim from 1911 was not justified and that his "celtium spectrum" was in fact due to element 71.[17] Then, in the spring of 1923, Urbain and Dauvillier reversed their stance concerning the chemical nature of element 72. While they had previously argued that it was a rare earth only accidentally found in zirconium minerals, they now recognized that the element should properly be counted as a homologue of zirconium. However, this change of mind seems to have been merely tactical, since the French team was not yet prepared to give up the claim for celtium.

The beginning of the hafnium–celtium controversy was most unwelcome to Bohr, who, being director of the Copenhagen institute where Coster and Hevesy worked, could not help but become involved in the conflict. To Rutherford, who submitted the Copenhageners' articles to *Nature,* Bohr expressed his dislike for

> this terrible muddle about the new element, in which we quite innocently have dropped.[18]

He continued,

> We had never dreamt of any competition with chemists in the hunt for new elements, but wished only to prove the correctness of the theory. In the letter from Urbain which the editor of Nature kindly sent us for possible comments, he tries however to shift the whole matter, paying no regard to the important scientific discussion of the properties of the element 72, but tries only to claim a priority for announcing a detection of such an element.[19]

Rutherford, who from the very start gave hafnium his full support, was equally tired of the debate, which he felt should end with Coster and Hevesy's criticism. "I quite agree with you that U. has not a leg to stand on," he commented on the Copenhagen article of February 9th, calling it "a document of an explosive character."[20]

The Role of Hafnium in Philosophy of Chemistry

Starting in the early 1990s, the philosophy of chemistry has become an area of active study within the philosophy of science. One of the major questions in this new subdiscipline has been whether chemistry is reduced to quantum mechanics, or the extent to which it may have been reduced.[21] Given the claims that have been made for the prediction by Bohr's theory that hafnium would be a transition element and not a rare earth element, and given the subsequent confirmation of this prediction by his colleagues Coster and Hevesy, a good deal of attention has naturally fallen on a close examination of this case. Moreover, the resolution of the priority conflict in favor of the Copenhageners and Bohr's theory led many outsiders to the field to make statements supporting the view that the hafnium episode meant that chemistry had been "reduced" to physics. For example, philosopher of science Karl Popper wrote,

> I still remember vividly the excitement of the discovery of the element 72 (Hafnium) in 1922, as a result of Niels Bohr's marvelous quantum theory of the periodic system of elements. It struck us then as the great moment when chemistry had been reduced to atomic theory; and it was, I am still inclined to say, the greatest moment in all the reductionist adventures of the twentieth century, superseded perhaps only by the breakthrough represented by Crick and Watson's discovery of the structure of DNA. Bohr's theory led not only to the prediction of the chemical properties of elements, and thereby to the prediction of the properties of the still unknown

element 72 and thus to its discovery, but it also allowed the prediction of some of their optical properties; and it even led to the prediction of some of the properties of the chemical compounds.

It was a great moment in the history of matter. We felt, rightly, this was it: Bohr had hit rock bottom. And yet a quite different type of problem was already looming in the background, started by a suggestion of Soddy's (1910) and a discovery of J. J. Thomson (1913) the year of Bohr's atom model, and by F. W. Aston's mass spectroscopy (1919). And then came Urey's bombshell, the discovery of heavy water, which meant that all the basic measurements of the atomic weights—the basic phenomena of chemistry and of the periodic system—were slightly wrong, and had to be revised. Thus the rock bottom suddenly gave way: somehow Niels Bohr had built on a morass. But this edifice still stood. Then came quantum mechanics, and the theory of London and Heitler. And it became clear that reduction of chemistry to physics was a reduction in principle only; and that anything like a complete reduction was now further distant than it had seemed in 1922, the year of the great breakthrough.[22]

Something of a new debate has begun to take shape some seventy years after the discovery of hafnium as to what role this episode should play in the question of whether quantum mechanics can provide a deductive explanation for chemical phenomena, or in this case the chemical nature of a particular element—hafnium.[23] In a paper published in 1994, I claimed that the conventional account of the discovery of hafnium was incorrect in one important respect, namely Bohr's role in directing Coster and Hevesy to look for signs of element 72 in the ores of zirconium. I claimed that the suggestion had in fact come from Fritz Paneth, the Berlin radiochemist who is frequently cited in other accounts but who is seldom credited with being the instigator of Coster and Hevesy's research on zirconium ores.

The usual accounts, as mentioned above, are that Bohr was so confident of his prediction that he directed Coster and Hevesy as to where to search for the new element. The main motivation for my questioning of the usual account was that I carried out my PhD in the history and philosophy of science with the son of Paneth, who provided me with correspondence belonging to his father that pointed to his much more important role in the episode than has been generally acknowledged.

In fact, in a number of earlier papers, Danish historian of science Helge Kragh had begun to raise some doubts. He too stated that the actual suggestion as to where to search had come from Paneth and not from Bohr himself. In addition, he raised what is perhaps a more important point in the question of the reduction or otherwise of chemistry. Kragh has argued, correctly I believe, that Bohr's predictions as to the chemical nature of hafnium were not as deductive as Bohr might have made them seem to be at the time. In addition, many other authors had noted that the electronic configurations that Bohr had assigned to the atoms of elements in the periodic table had, in fact, been arrived at in a semiempirical manner and not from strict deductions from his quantum theory.[24]

> This which is very characteristic of Bohr's approach, thus consists of two rather distinct elements, namely an application of general theoretical principles and an application of chemical or physical knowledge. This division, which was a general trend in atomic theory of that time and which was characteristic for the 1921 theory also, corresponds to reasonings of, respectively and deductive and an inductive character. In Bohr's work, where the mechanical and quantum theoretical principles could only fix the configurations very grossly, the chemical-inductive arguments played the major role.

It has been a little surprising to me that Kragh has several times objected to my raising concern over the extent to which Bohr really predicted the chemical nature of hafnium. For example, in a footnote to an article on the notion of an element, Kragh writes,

> Eric Scerri (1994) has argued that Bohr's theory did not predict the chemical nature of element no. 72 and that the discovery can

> therefore not be regarded a triumph of the theory. It is correct that Bohr's prediction was not very conclusive and that the insight of the element being a zirconium homologue was occasionally reached from chemical reasoning alone. However, from a historical point of view Scerri's objections are irrelevant. In any reasonable sense of the term, Bohr predicted the chemical properties of hafnium, although not in a strict deductive manner. Almost all physicists accepted the predictive link between theory and discovery and considered the discovery of hafnium a great triumph of Bohr's atomic theory.[25]

Perhaps the disagreement comes from the fact that Kragh is approaching the subject from the perspective of physics whereas I approach the matter from a predominantly chemical perspective.[26]

Hafnium. Properties and Applications

The chemical properties of hafnium are extremely similar to those of zirconium, which lies above it in group IV of the periodic table. For example, the ionic radius of Zr is 0.74 Ångstroms, compared to 0.75 Ångstroms in the case of hafnium, a difference of just 1 percent. This occurs despite the fact that the nucleus of hafnium contains an additional 32 protons when compared to the nucleus of zirconium, not to mention many additional neutrons.

The effect arises partly due to the "lanthanide contraction." As one crosses the third transition series of elements, the size of atoms are smaller than expected because the additional electrons, added to the atom as the series is traversed, fall into an inner f orbital (fig. 4.5). These 4f orbitals represent the ante-penultimate shell and provide poor shielding of outer electrons as compared to shielding by d-electrons, for example. As a result of this poorer shielding, the nuclear attraction is relatively greater and most of the elements in the third transition series have atomic and ionic radii that are close to the sizes of the corresponding atoms in the second transition series.

La	Ce	Pr	Nd	Pm	Eu	Gd	Tb	Dy	Ho	Er	Tm	Yb
1.00	1.02	0.99	0.983	0.97	0.958	0.938	0.923	0.912	0.901	0.890	0.880	0.868

FIGURE 4.5 Variation in ionic radii in Å, showing the lanthanide contraction.

As a consequence of this and many other similarities, atoms, and especially the ions of hafnium and zirconium, can easily substitute for each other in the minerals in which they are invariably found together. There is no chemical means of separating the two elements since any separation needs to exploit differences in behavior and the differences are almost insignificant.

On the other hand, there are some physical differences between the two elements. For example, the densities of zirconium and hafnium are rather distinct, with zirconium being only half as dense as hafnium. Their melting points are also quite different. Of the 34 presently known isotopes of hafnium, the least stable is ^{153}Hf, with a half-life of 400 milliseconds, while ^{174}Hf has a half-life of 2×10^{15} years.

Perhaps the major difference between the two elements is their neutron-absorbing power. Hafnium can absorb neutrons about 600 times as well as zirconium can. As a result, the main application of hafnium is in nuclear reactors in the form of control rods. Moreover, this use has meant that great efforts have been made to improve the methods for separating the two elements, in order to maximize the neutron-absorbing tendency of hafnium.

Hafnium holds some records concerning high melting points. For example, hafnium carbide, HfC, is the single highest melting point binary compound that is known, with a value of more than 3980°C. In addition, the mixed carbide with a formula of Ta_4HfC_5 has the highest melting point of *any* compound known.

Other uses include making alloys with metals such as titanium, niobium, and tantalum. For example, the nozzles in rocket thrusters such as the Apollo Lunar Modules were made of an alloy of niobium, hafnium, and titanium.

Chapter 5

Element 75—Rhenium

H																	He
Li	Be											B	C	N	O	F	Ne
Na	Mg											Al	Si	P	S	Cl	Ar
K	Ca	Sc	Ti	V	Cr	Mn	Fe	Co	Ni	Cu	Zn	Ga	Ge	As	Se	Br	Kr
Rb	Sr	Y	Zr	Nb	Mo	Tc	Ru	Rh	Pd	Ag	Cd	In	Sn	Sb	Te	I	Xe
Cs	Ba	Lu	Hf	Ta	W	**Re**	Os	Ir	Pt	Au	Hg	Tl	Pb	Bi	Po	At	Rn
Fr	Ra	Lr	Rf	Db	Sg	Bh	Hs	Mt	Ds	Rg	Cn		Fl		Lv		

La	Ce	Pr	Nd	Pm	Sm	Eu	Gd	Tb	Dy	Ho	Er	Tm	Yb
Ac	Th	Pa	U	Np	Pu	Am	Cm	Bk	Cf	Es	Fm	Md	No

FIGURE 5.1 Showing the position of element 75, eventually named rhenium in the periodic table.

The element rhenium lies two places below manganese in group VII of the periodic table (fig. 5.1). Its existence was predicted by Mendeleev when he first proposed his periodic table in 1869. This group is rather unique because when the periodic table was first published, it possessed only one known element, manganese, with at least two gaps below it. The first gap was eventually filled by element 43, technetium, while the second gap was filled by rhenium. But rhenium was the first to be discovered, in 1925, by Walter Noddack and Ida Tacke (later Noddack) (fig. 5.2) and Otto Berg in Germany.[1] In the course of an arduously long extraction, they

FIGURE 5.2 Ida Noddack (née Tacke). Image by permission of Emilio Segrè Collection at the Institute of Physics.

obtained just one gram of rhenium after processing about 660 kg of the ore molybdenite.[2]

The German discoverers called their element "rhenium" after Rhenus, Latin for the river Rhine, which flowed close to the place where they were working. They also believed that they had isolated the other element missing from group 7, or element 43, which eventually became known as technetium, but this was hotly disputed by several other researchers.

As recently as the early years of the twenty-first century, research teams from Belgium and the United States reanalyzed the X-ray evidence from the Noddacks and argued that they *had* in fact isolated element 43.[3] But these further claims have been debated by many radiochemists and physicists and now have been laid to rest, at least for the time being.[4]

By a further odd twist of fate, the Japanese chemist Masataka Ogawa believed that he had isolated element 43 and called it nipponium even earlier, in 1908. His claim too was discredited at the time but as recently as 2004 it has been argued that he had in fact isolated rhenium rather than element 43, well before the Noddacks and Berg.[5]

Nipponium

Otto Hahn's first entry into the field of radioactivity was as a student of Ramsay's at University College, London, just after the beginning of the twentieth century. There, Ramsay asked him to analyze the ore thorianite, which had been especially imported from Ceylon (now Sri Lanka). As a result, Hahn was able to make a very early impact on the field, initially believing that he had discovered a new element that later turned out to be a new isotope of thorium. His mentor, Ramsay, had been the codiscoverer, with Rayleigh, of several new elements (neon, argon, krypton, and xenon), the noble gases that eventually necessitated the creation of an entirely new group in the periodic table.

By something of a coincidence, Ramsay also asked another young student to analyze the same ore, presumably in another bid to find a new element. The student was Masataka Ogawa, who arrived at University College from Japan in 1904. After a good deal of analytical work, Ogawa thought he had isolated a new element and convinced Ramsay that he had done so. Among the evidence put forward by Ogawa was the detection of a spectral line at 4882 +/- 10 Å.[6] He calculated the atomic weight of the element to be 100 based on the belief that it was divalent.[7]

Ramsay seems to have approved of the finding and even suggested to Ogawa that it should be named nipponium, after the Japanese name for Japan. On returning to Japan, Ogawa reported that he had found evidence for this new element in molybdenite[8] and published two articles in the London-based journal *Chemical*

News.[9] A periodic table published the following year by Loring, also in *Chemical News,* accepted the new element, which was given the symbol of Np and placed in the position that today is occupied by element 43, or technetium.

Unfortunately, further efforts to replicate these findings by younger colleagues of Ogawa, at his university of Tohoku, were unsuccessful. This can be partly explained by the fact that thorianite, which the young Japanese colleagues were working with, contains very little rhenium. More precisely, 96.9 percent of the mineral is made up of ten elements, of which thorium is of course the major component. The remaining 3.1 percent consists of a further eleven elements, of which rhenium is just one minor portion. Whereas Ogawa's original experiments in London had been conducted on 250 kg that were made available to him, the later Japanese researchers were working with less than a kilogram of the mineral.

A number of other researchers also claimed to have found eka-manganese, or element 43. They included F. H. Loring and J. G. F. Druce in England who claimed that they had extracted the new element from manganese salts and pyrolusite (MnO_2) using electrochemical methods. Some others also searched for eka-manganese and admitted that they could not find it, including O. Zvjagincev in Russia. Another chemist, W. Prandtl in Germany, searched in columbite, tantalite, gadolinite, and wolframite, all to no avail.

Noddack, Tacke, and Berg

In 1925, Noddack, Tacke, and Berg, working in Germany, reported that they had discovered two new elements in group seven, which they named masurium and rhenium, respectively. By an odd twist of fate, their claim for rhenium was upheld whereas that for technetium was not. Meanwhile, Ogawa's nipponium disappeared almost without a trace, although it was later resuscitated by Yoshihara, who claimed that Ogawa had, in fact, discovered rhenium and not element 43 as he thought he had. In fact, element 43 eluded Ogawa as well as the Noddacks and was only isolated, or rather synthesized, by Perrier and Segrè in 1937.[10]

The Noddacks went beyond their competitors because they fully realized that the two elements they were seeking were not like manganese but more akin to their horizontal partners. For example, whereas manganese disulphide is soluble in acids, the Noddacks believed that elements 43 and 75 would be insoluble. They began by eliminating iron and manganese from their crude ores by precipitation and filtration of these two metals. This involved making a total of 400 enriched products from different ores. In order to confirm the identity of the new elements, they enlisted the help of Otto Berg, an X-ray specialist at the Siemens-Halske Company in Berlin. In June 1925, the Noddacks along with Berg announced that they had identified a new element they proposed to call rhenium, from a Norwegian columbite ore.

With the aid of a 30,000 Reich Mark grant from the German Scientific Energy Fund, the Noddacks traveled to Scandinavia and Russia in order to purchase further minerals they believed might contain rhenium. Their first success came in 1927 when they obtained 120 mg of rhenium and studied some of its chemical properties. There was commercial interest in the new metal almost immediately. Later in the same year, Siemens and Haske, where Berg worked, contracted the Noddacks to extract 1 gram of rhenium. The conditions of this contract were that the extracted metal would become the property of the company, while still allowing the Noddacks to conduct any further experiments. By 1929, the Noddacks duly delivered a whole gram of the metal following extraction from 660 kg of Norwegian molybdenite, reaching a mass of 3 grams after just one more year.

Further analysis showed that rhenium is present in the earth's crust at a very low concentration of 10^{-7} percent, or 0.01 p.p.m. Although it does not form its own particular minerals, rhenium occurs in molybdenite as has already been mentioned, and also in porphyr copper ores. The metal has a melting point of 3180°C, making it only second to tungsten, which melts at 3380°C. The specific gravity of rhenium is the fourth highest of any element following Os, Ir, and Pt.

Applications of Rhenium

Let's turn to the chemistry and properties of rhenium. Until quite recently, no mineral containing rhenium combined with just a nonmetal had ever been found. However, as reported in *Nature* in 1992, a team of Russian scientists discovered rhenium disulphide at the mouth of a volcano on a remote island off the east coast of Russia.[11] The volcano, called Kudriavy, has not erupted since 1883 but contains gas jets that reach a temperature of over 900°C. By subliming these gas jets, the Russian team discovered what they call "the first reported pure mineral of rhenium," and one that contains rhenium as the only cation.

The chemistry of rhenium is rather diverse. Among other things, it shows the largest range of oxidation states of absolutely any known element, namely −1, 0, +1, +2, and so on all the way to +7, the last of which is its most common oxidation state. It is also the metal that led to the discovery of the first metal-to-metal quadruple bond. In 1964, Albert Cotton and coworkers in the United States discovered the existence of such a Re–Re quadruple bond in the form of the rhenium ion, $[Re_2Cl_8]^{2-}$.[12]

A large quantity of extracted rhenium is made into superalloys to be used for parts in jet engines. Typically for a transition metal, rhenium also acts as a catalyst for many reactions. For example, a combination of rhenium and platinum make up the catalyst of choice in the very important process of making lead-free and high-octane gas. Rhenium catalysts are especially resistant to chemical attack from nitrogen, phosphorus, and sulfur, which also makes them useful in hydrogenation reactions in various industrial processes.

More recently, a rather simple compound of the element, rhenium dibromide, has attracted some attention because it is one of the hardest of all known substances. Unlike other super-hard materials like diamond, it does not have to be manufactured under high pressure.[13]

Modern Applications of Rhenium

In modern industry there is a large need for hard materials and also for so-called super-hard materials that are used as abrasives, cutting tools, and scratch-resistant coatings. The role of a super-hard material has traditionally been played by diamond, with its well-known ability to act as a cutting tool. Other super-hard materials have also been developed as substitutes, including cubic boron nitride, or BN. However, such diamond substitutes necessitate the use of extremely high pressures to produce them, typically more than five giga pascals and temperatures in excess of 1,500°C, all of which renders manufacture rather costly.

In 1962, La Placa and Post, working in New York, stumbled upon the synthesis of a new compound of rhenium, namely the exceedingly simple diboride with formula ReB_2.[14] But they did not realize the potential of this compound as a super-hard material. It was left to two of my UCLA Chemistry Department colleagues, Richard Kaner and Sarah Tolbert, to appreciate this possibility and carry out tests and characterizations that have firmly established that the compound is indeed super-hard.[15]

The two factors that give rhenium diboride its hardness are a high density of valence electrons and an abundance of short covalent bonds. Kaner and Tolbert had previously singled out osmium diboride as a candidate super-hard material and had shown that it had many desired characteristics. They then consulted the periodic table in order to find other elements that might provide greater hardness still. Rhenium lies one place to the left of osmium in the table and is therefore an obvious choice. Microindentation studies and the fact that rhenium boride leaves a scratch on the surface of a diamond have contributed to the argument that ReB_2 is a super-hard material and have opened up the possibilities of technological applications for this compound.

For example, rhenium boride can be used as a cutting tool in cases in which diamond is problematic because of its tendency to form carbides. This happens when diamond is used to cut iron or steel, especially in cases of high-speed machining.

Rhenium is the last stable element to be discovered but is certainly not insignificant. It has many interesting and useful properties, such as the fact that it does not undergo a ductile-brittle transformation as its melting point is approached, in contrast to many other metals. Rhenium retains a very high strength at high temperatures in addition to very good ductility, making it an ideal choice for high temperature applications. One of these applications is in the making of aircraft turbine blades because the addition of 1–3 percent rhenium to a nickel alloy improves its toughness at high temperatures and prevents fatigue fractures.

Other applications of rhenium alloys include the making of semiconductors, nuclear reactors, electronic tube components, thermocouples, electrical contacts, gyroscopes, and X-ray tubes. The metal is also widely used in the chemical industry as a catalyst to bring about alkylation, dealkylation, dehydrochlorination, hydrocracking, oxidation, and reforming. One of its attributes is a resistance to some typical catalyst poisons such as nitrogen, sulfur, and phosphorus. The world's largest reserves of rhenium are found in Chile, which is also the biggest producer of the metal.[16]

Some Contemporary Rhenium Research

Dean Harman at the University of Virginia has been pioneering the use of rhenium compounds in basic synthetic organic chemistry.[17]

The work began when Harman was still a student with Nobel laureate Henry Taube at Stanford, where they discovered the important role of a transition metal compound, pentaammineosmium(II) or $[Os(NH_3)_5]^{2+}$. When this fragment molecule binds to benzene rings and similar aromatic compounds, it causes dramatic changes in their behavior and gives them promising new properties from the point of view of synthetic organic chemistry. Whereas aromatic compounds are typically susceptible to substitution reactions, they now become prone to addition reactions. This change occurs because the aromatic behavior of the cyclic compounds has been reduced via a process called dearomatization. Since the mid-1980s, various other such compounds were synthesized, including transition metal complexes of Cu(I), Ag(I), Pt(0), Ni(0), Ta(III), Nb(III), Rh(II), and so on.

In many of these compounds, a transition metal, such as Re, can form a single bond with the aromatic compound, resulting in the other two "double bonds"—thinking in terms of classical organic chemistry—becoming localized to produce a compound of cyclohexadiene. This process is known as dearomatization, since it results in the loss of the aromatic, or delocalized, nature of the benzene ring (fig. 5.3).

Because the aromatic stabilization of the original aromatic compound is thereby broken, many organic reactions that would otherwise not be possible can now be performed. Organic synthetic chemists have thus gained some valuable new tools in their arsenal of methods.

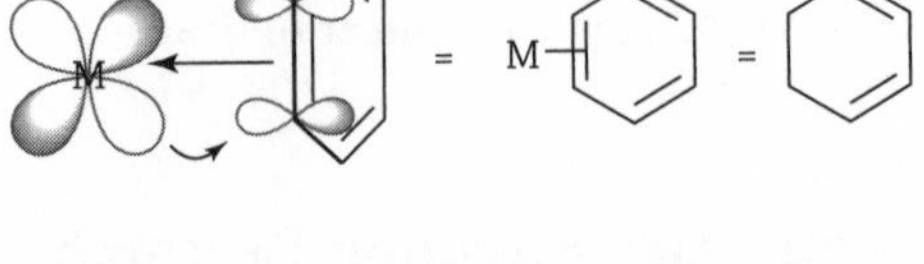

FIGURE 5.3 Dearomatization reaction using metal atoms. J. M. Keane, W. D. Harman, A New Generation of π- Basic Dearomatization Agents, *Organometallics,* 24, 1786–1798, 2005. Reprinted with permission of the American Chemical Society.

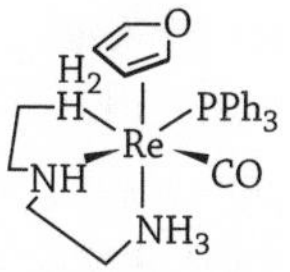

FIGURE 5.4 Second generation dearomatization agent featuring rhenium atom. J. M. Keane W. D. Harman, A New Generation of π-Basic Dearomatization Agents, *Organometallics,* 24, 1786-1798, 2005. Reprinted with permission of the American Chemical Society.

Even more recently, a so-called second generation of dearomatization agents have been discovered by focusing on low valence states of the metal rhenium. One such compound, whose formula would take up half a line of text, is shown in figure 5.4, above.

An Attempt to Rehabilitate Nipponium

Over an extended period of time, the Japanese emeritus professor Kenji Yoshihara has succeeded in publishing a number of articles in which he proposes to rehabilitate Ogawa's claim to having discovered a new element, called nipponium.[18] Given the recent and very persistent claim for the "true discovery" of one of our seven elements, it will be examined in some detail.

As mentioned earlier, the Japanese chemist Masataka Ogawa went to London to study with William Ramsay in 1904. Ogawa was given a sample of thorianite that had come from Ceylon (now Sri Lanka) and asked to explore its contents. In 1908, Ogawa reported seeing what he believed was a new spectral line in the optical region that had not been previously assigned to any element and that occurred at a wavelength of 4882 Å. Upon hearing of this work, Ramsay became convinced that his Japanese student had discovered a new element and suggested to Ogawa that it should be called nipponium.

Ogawa calculated the atomic weight of the "new element" to lie between those of molybdenum and ruthenium. This would place it as lying directly below manganese, as predicted by Mendeleev, who gave it the name eka-manganese. Ogawa returned to Japan in 1906 and continued searching for this element, claiming later that he had also found it in Japanese molybdenite. He published two articles in Crookes's London journal *Chemical News*[19] and the symbol Np for nipponium promptly appeared in a periodic table published by Loring.

In 1911, Ogawa moved from the University of Tokyo to Sendai University, where he became president in 1919. His colleagues and students tried to repeat the detection of nipponium but failed to do so. Meanwhile, Hevesy, the discoverer of hafnium, was sent a sample of Ogawa's purported nipponium-bearing ore but also failed to observe any new element. Writing in French in 1925, Hevesy stated:

> Quelques années plus tard, OGAWA cruit avoir trouvé, dans le thorianite, le silicate d'un élément nouveau, le nipponium. M. R.-B. Moore, chimiste principale du Bureau de Mines à Washington, a eu l'extrême obligeance de nous addresser quelque-uns des cristaux de silicate de nipponium obtenus par Ogawa. Ces criteaux se composaient, essentiellement, de silicate de zirconium ayant une teneur de 2 p.c. en hafnium.[20]

Subsequent authors on the periodic table, such as van Spronsen, all accepted this negative finding. The claim appears to have persisted in Japan, however, especially in light of still-living nuclear chemist Kenji Yoshihara, who has devoted much time to trying to restore Ogawa's claim while also concluding that it was wrongly assigned as eka-manganese and should have been attributed to the element below this space, which was eventually called rhenium.

Yoshihara does not dispute the veracity of this sample that was sent to Hevesy but conducts a detailed examination of Ogawa's method of separation to conclude that:

> In the above procedures [Ogawa's], the separated sample was nearly free from zirconium and silicate contrary to the description of Hevesy. Could Ogawa mistake zirconium as nipponium even if a small quantity of zirconium contaminated the final fraction as obtained in Fig 2? It was unthinkable for Ogawa as an expert of separation experiments! p. 264.

Needless to say, such a statement made on what appears to be somewhat nationalistic grounds is hardly compelling in scientific terms. Neither is the following statement by Yoshihara:

> Considering that Ogawa looked for the element for [all] his life under the strong support of Ramsay, it is strange why he worked on such an objective as a phantom. What was nipponium in reality? The question was unanswered until recently. p. 265.

Yoshihara finally turns to a more substantial argument in saying,

> The key to solve the question was the spectroscopic observation that Ogawa described in his report. A new spectral line of 4882 Å (with experimental error of 10 Å) emitted from the "new element" is in good agreement with that of rhenium (4889 Å) . . . Therefore Ogawa's nipponium should be assigned to rhenium spectroscopically. p. 265.

Yoshihara also claims that Ogawa's calculated atomic weight for nipponium was incorrect because Ogawa believed that the element was divalent. Yoshihara's recalculated atomic weight, assuming a hexavalency for the element, reveals an atomic weight of 185.2, which he claims is very close to the modern-day value of rhenium: 186.2. Yoshihara explains to the reader that he presented his findings to an international conference held in Belgium in 1996 and that it "gave a strong impact on the people interested in the history of chemistry in Europe and the United States." He then adds that he published a further paper in the journal *Radiochimica Acta*, which "was highly evaluated by many people."[21]

Presumably realizing the tenuous nature of his claims, Yoshihara finally plays his trump card. He reports that Ogawa had always wanted to obtain X-ray confirmation of his discovery but that no adequate X-ray machine has existed in Japan.[22] Then, in 1927, Kenjiro Kimura purchased an adequate X-ray spectrometer and indeed attempted to verify Ogawa's new element. Unfortunately, no paper was ever published to report the results of this analysis but Yoshihara informs us that yet another Japanese professor, Toshi Inoue, "who was a very good friend of Kenjiro Kimura," is supposed to have exclaimed,

> It was beautiful rhenium, indeed!

In a paper written in 2005, Yoshihara reports what he calls new evidence. It appears that the x-ray plates from Kimura's experiments have been found. Not the Kimura plates, unfortunately, but those of yet another character in the unfolding story, a Professor Aoyama. After much decipherment and speculation, Yoshihara interprets them to show that there was in fact no element 43 in Ogawa's sample but that there are two sharp peaks (fig. 5.5) that can be

> clearly assigned to those of $L_{\beta1}$ and $L_{\beta2}$ X-rays of rhenium.

At first sight this seems to be the best evidence that Yoshihara has produced to date, but this too leaves a number of questions unanswered. First of all, the two curves do not consist of the X-ray lines obtained by Aoyama and those obtained on "real" rhenium as one would have hoped for. They are a comparison of Aoyama's peaks with the peaks calculated for rhenium.[23]

My own conclusion is that there may well have been some rhenium in Ogawa's samples for the simple reason that rhenium is not such a rare element. But this is quite a different matter from claiming that Ogawa's samples did actually contain rhenium—something that I do

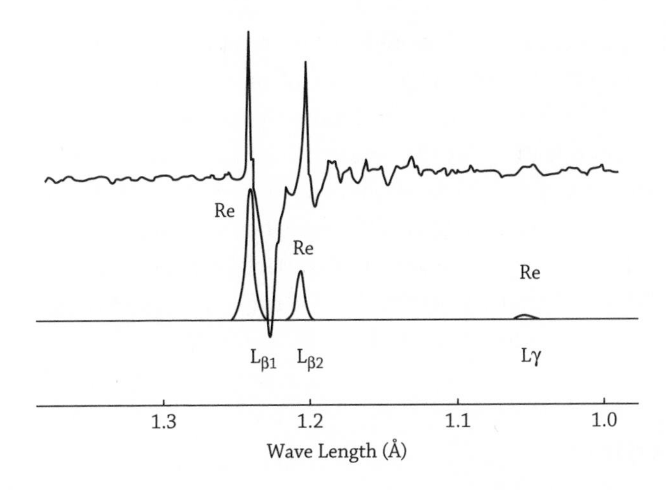

FIGURE 5.5 X-ray peaks from Kimura's experiments as reported by Yoshihara in 2004. Yoshihara, H. K. Discovery of a new element nipponium: reevaluation of pioneering works of Masataka Ogawa and his son Eijiro Ogawa, *Spectrochimica Acta Part B Atomic Spectroscopy 59*, 1305–1310, 2004. Reprinted with permission of Elsevier.

not believe has been established either by the circumstantial evidence on how well thought of Ogawa might have been or how well the papers that Yoshihara has subsequently written might have been received.

One final point from Yoshihara's articles should also be considered. In a number of instances, this author has maintained that the Japanese researchers who validated Ogawa's claim were prevented from announcing their results because of a Japanese tradition whereby more senior professors should not be contradicted by their juniors. Yoshihara is alluding to the fact that these subsequent researchers had realized that Ogawa's initial claim to having found element 43 was incorrect. He is contending that drawing attention to this fact would have represented a form of insubordination at the time when these events were unfolding.

I must say that I find this further claim implausible as well, since any announcement that Ogawa had indeed discovered rhenium, rather than element 43, would still have represented an important development and would have brought prestige to Ogawa. Such an announcement would have undoubtedly compensated for any mistaken identification on Ogawa's part. I cannot believe that it was the reason why several researchers failed to publish their claim that Ogawa had, in fact, isolated what later became known as rhenium. The actual explanation is far simpler. Back then, as today, there was insufficient evidence to claim that Ogawa's nipponium was, in fact, none other than the element rhenium.

Interlude

By 1925, all but one of the elements that occur naturally had been discovered. Just four gaps between the old limits of the periodic table remained to be filled. Although there were numerous claims for these missing elements, which were given fanciful names including,

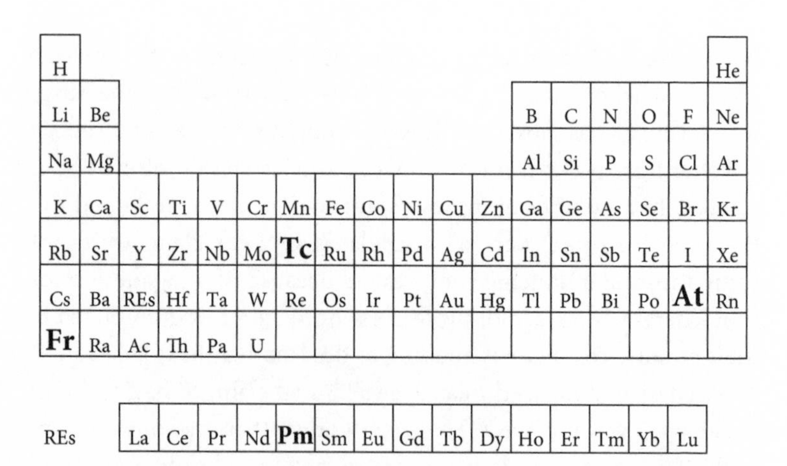

FIGURE 5.6 Positions of the four elements, which were yet to be discovered after the detection of rhenium in 1925 (shown in bold letters) on a medium-long form periodic table of the time.

masurium, illinium, florentium, alabamine, virginium, moldavium, russium, and a host of others, none of these claims have stood the test of time. It is also fairly clear that none of these claims could possibly have been correct given the highly unstable nature of the elements and the fact that, with the exception of francium, they had to be synthesized rather than discovered among naturally occurring sources.

The last four remaining elements—eventually named technetium, astatine, francium, and promethium—therefore form a separate, though not too well-defined, subclass in the tale of the seven elements (fig. 5.6). They can also be regarded as forming a separate group of elements in view of the relatively long time delay before they began to be discovered. From the time of the discovery of hafnium in 1925, it was another twelve years before the next element, the first of our final four elements, was eventually synthesized in 1937. It took a further ten years, however, before the name "technetium," as suggested by its discoverers, became accepted.

Final note: As this book was going to press Oliver Sacks reminded me of the claim made in 1869, the year of Mendeleev's table for an element called jargonium. This element which was thought to occur with zirconium was announced by H.C. Sorby in *Chemical News*, only to be withdrawn a year later for lack of sufficient evidence.

Chapter 6

Element 43—Technetium

H																	He
Li	Be											B	C	N	O	F	Ne
Na	Mg											Al	Si	P	S	Cl	Ar
K	Ca	Sc	Ti	V	Cr	Mn	Fe	Co	Ni	Cu	Zn	Ga	Ge	As	Se	Br	Kr
Rb	Sr	Y	Zr	Nb	Mo	**Tc**	Ru	Rh	Pd	Ag	Cd	In	Sn	Sb	Te	I	Xe
Cs	Ba	Lu	Hf	Ta	W	Re	Os	Ir	Pt	Au	Hg	Tl	Pb	Bi	Po	At	Rn
Fr	Ra	Lr	Rf	Db	Sg	Bh	Hs	Mt	Ds	Rg	Cn		Fl		Lv		

La	Ce	Pr	Nd	Pm	Sm	Eu	Gd	Tb	Dy	Ho	Er	Tm	Yb
Ac	Th	Pa	U	Np	Pu	Am	Cm	Bk	Cf	Es	Fm	Md	No

FIGURE 6.1 The position of technetium in the periodic table.

Element 43 (fig. 6.1) holds a special distinction among the seven elements of this book. It was one of just four elements that Mendeleev first predicted in his famous table and article of 1871. This fact is not so well known, as most accounts mention just the three famous predictions, namely empty spaces to which Mendeleev gave atomic weights of 44, 68, and 72. These three elements were all discovered within a period of fifteen years and named scandium, gallium, and germanium, respectively. But in the same early table, Mendeleev assigned an atomic weight to just one more empty space, which he placed below manganese. Mendeleev predicted that it would have an atomic weight of 100, although he changed it slightly to 99 in his book, *The Principles of Chemistry*.[1]

Given the success of Mendeleev's first three predictions it is hardly surprising that strenuous efforts were made, in many parts of the world, to find the fourth element. Little did these early chemists know the problems they would encounter in trying to isolate this particularly rare and unstable element. In the early twentieth century, several claims were made for the discovery of the element. But these alleged elements, given various names such as davyum, illenium, lucium, and nipponium all turned out to be spurious. Then, in 1925, as mentioned in the last chapter, Otto Berg, Walter Noddack, and Ida Tacke (later Ida Noddack), claimed to have discovered not just one but two new members of group 7, which they named masurium and rhenium. Although their discovery of rhenium was accepted, their claim for the element directly below manganese has been bitterly disputed ever since.[2]

The official discovery of element 43 is accorded to Emilio Segrè and coworkers. Technetium, as they eventually called it, had to be synthesized rather than isolated from naturally occurring sources. It is also the only element to ever be "discovered" in Italy—in Palermo, Sicily, to be more precise. Segrè, who had been a visitor at the Berkeley cyclotron facility in California, was sent some molybdenum plates that had been irradiated for several months with a deuterium beam. Various chemical analyses by the Italian team revealed a new element, which could be extracted by boiling with sodium hydroxide that also contained a small amount of hydrogen peroxide.

It is generally believed that any technetium that might have been present when the earth was first formed has long since decayed radioactively. We know this because even the longest-lived isotope of the element has a half-life that is too short in comparison with the age of the Earth. But in 1956, the Japanese radiochemist Paul Kuroda predicted that a natural nuclear reactor might once have existed deep within the Earth.[3] Five years later, he reported that a sample of African pitchblende contained about 2×10^{-10} grams of ^{99}Tc per kilogram of ore. In 1962, a team of French scientists confirmed Kuroda's earlier prediction of a natural nuclear reactor on investigating rock samples in the Republic of Gabon in Africa.[4] Further analyses showed that there were trace amounts of technetium present in these minerals too, thus further contradicting the common textbook statement that technetium does not occur naturally on Earth.

Early Claims for Element 43

Rather remarkably, many claims for the discovery of the element that in modern terms would be regarded as element 43 were made in the early and mid-1800s. Needless to say, the concept of atomic number was not known at this time, which means that any claims would not have referred to element 43 as such. Moreover, no periodic tables had yet been published to guide such research and nobody had even thought of arranging the elements in any particular order to see what gaps might have existed. The very early claims that will first be described were thus made entirely on the basis of chemical similarities with elements that were known to exist.

Very Early Claims

Several authors have identified the very first such claim with an element that was given the name polinium. The story begins in 1823, when the imperial government of Russia decreed that any samples of platinum ores should be sent to St. Petersburg for chemical analysis. This was followed two years later by the Russian state declaring a monopoly on the metal.[5]

A chemist named Egor Kankrin followed these orders, to a certain extent, by sending samples of platinum-bearing ores to St. Petersburg. He did not, however, restrict himself to only Russian institutions. He also sent samples to the Swedish chemist Berzelius and to one Gottfried Wilhelm Osann, a professor of chemistry in what is now Tartu in modern-day Estonia. In 1927, Osann announced that he had discovered three new elements, including one that he called polinium (from the Greek πολιοσ, meaning gray).

He named the other two new suspected elements, ruthenium (after Russia) and pluranium (because it was present to a larger extent than Osann's polinium).

> I have discovered in the platina of Uralian mountains three metals, the properties of which are different from those of every other known metal. One of them occurs in the residium left by the solution of the platina in aqua regia, which is sold at the Mint in Petersburg. I have as yet found it only in one specimen of the metal.

Osann sent samples of his three new elements to Berzelius, asking for his opinion. Berzelius responded by dismissing the oxide of Osann's ruthenium and finding that it was in fact a mixture of silicon, zirconium, and titanium. On the other hand, he found that the oxide of Osann's pluranium seemed to be a new substance. As to polinium, Berzelius wrote to Osann that he could not decide upon its true nature.

In 1831, Karl Ernst Klaus, a self-taught chemist, was appointed to a junior post in the same university as Osann, and learned of these suspected new elements. By 1840, Klaus had become a professor of chemistry in Kazan, Russia. Two years later he met Kankrin, who had begun this chain of events back in the 1820s and who in turn provided him with eighteen pounds of residue ores from the Imperial Mint. While trying to extract platinum from this material, Klaus then succeeded in discovering a new element, which he announced by writing,

> These residues were poorer than the first... and my hope of... profitable extraction of platinum from them was not fulfilled. (However) I have worked constantly on this hard, prolonged and unhealthy investigation, and now I report to the scientific world... the discovery of a new metal, ruthenium... and new properties and

> compounds of the previously known metals of the platinum group. All this may serve as a contribution to the chemical history of a precious product of our fatherland.

After treating the residue ores with alkali and potassium nitrate and extracting with water, he removed the osmium present by various means, to reveal a red solution consisting of a chloride of the new metal. Klaus decided to call this metal ruthenium in honor of Osann, who had first coined the name, as well as in honor of his homeland of Russia.

He also expressed the view that Osann's ruthenium oxide must have contained the metal in low concentration. Klaus then sought the advice of Berzelius, as Osann had done. As Griffith writes in his article of 1968,

> Berzelius, no doubt bored with the whole affair by now, was at first skeptical and inclined to think that the new element was simply impure iridium, but later changed his mind and wrote again to Klaus, agreeing that ruthenium was a new element, and urging him to report his work in the German journals.

Klaus promptly did what he was told and the new metal was so announced in 1845, whereupon Osann wrote to the journal to claim that his polinium was identical with Klaus's ruthenium. However, Klaus responded politely that polinium had been no more than another impure metal, namely iridium. Osann's polinium was not a genuine element but as Griffith concludes, it was his work that led Klaus to discover ruthenium, a metal for which Osann's name was retained.

Ilmenium and Neptunium

The next element to be considered, like the first one, polinium, has a rather tenuous connection with element 43. Some authors in this field, like Kenna, and more recently, Zingales, consider it to be a plausible early candidate for what became element 43, whereas other authors do not.[6]

In 1846, Rudolph Hermann, in Germany, believed that he had discovered a new element among some ores of niobium and tantalum mined from the Ural Mountains, a claim that was soon criticized by others. Thirty years later Hermann revived his assertion once more and also claimed another new element that he named neptunium to occupy what had by now been coined dwi-manganese by Mendeleev. Both of these elements were shown to be spurious and nothing was heard of them again, except for the fact that neptunium was eventually used to name element 93.[7]

Davyum

In 1877, Sergius Kern, a Russian chemist working for the Obouchoff Steel Works in St. Petersburg, claimed to have perceived

> The presence of a new metal of the platinum group, which has been called by me Davyum, in honor of the great English chemist Sir Humphry Davy.

Kern reported that the new metal was soluble in aqua regia, showed at atomic weight of 154, and gave a red color when thiocyanate ions were added. That same year, however, A. H. Allen pointed out that Kern had often made arithmetical mistakes and that like his arithmetic, his chemical claims were not to be trusted. To add insult to injury, Allen added:

> It is evident that calculations are not Mr. Kern's strong point . . . unfortunately that gentleman's contributions contain little that is novel, and that little is mostly inaccurate.

Kern's work was also examined critically by J. W. Mallett from the University of Virginia. Although he succeeded in duplicating many of Kern's findings, Mallett's conclusion was that the presumed new metal was no more than a mixture of iridium and rhodium.[8]

Uralium

In 1869, the year that Mendeleev published his epochal periodic table, A. Guyard, in France, claimed to have found an element whose atomic weight of 187 and density of 20.25 were close to the values expected for dwi-manganese. On the other hand, the properties of this alleged element, which Guyard dubbed as uralium, were reported as being those of a brilliant white color and as soft as lead. The fact that rhenium turned out to be quite different in both respects suggests that uralium, too, should be confined to the wastebasket of elements that failed to materialize.

Canadium

Yet another spurious claim for Mendeleev's dwi-manganese was published by A. G. French in 1911 in a Glasgow newspaper and a little later in *Chemical News*. The discoverer called it canadium because he found the substance in British Columbia in Canada. Rather surprisingly, the claim was not accompanied by any scientific data, not even when it appeared in *Chemical News,* a feature that drew much criticism in a subsequent published note to the editor.

> Is it advisable to publish such unripe researches as those of Mr. French?...the whole investigation is done in such an unscientific manner that it does not inspire confidence at all, being as it is conducted businesslike and commercial lines.

This was because all of French's comments on his claimed new element had been directed at possible commercial exploitation and applications of the purported metal.[9]

Neo-molybdenum

This was the name proposed for element 43 by the French chemist Gerber writing in 1917 in the journal, *Le Moniteur Scientifique de Quesneville*.[10] His basis for doing so was that he believed that the missing elements 43 and 75 would show a greater resemblance to elements lying close to them in a horizontal direction of the periodic table rather than the more expected similarities within their own group. In this respect, Gerber would later turn out to be correct since the discovery of element 75 was achieved by the Noddacks' reasoning precisely in this manner, as we will see shortly.[11]

Moseleyum

In 1924, the chemists Bosanquet and Keeley initiated an extensive search for element 43 but took a step back in the wrong direction by searching in the ores of manganese. They believed that their strategy would be successful, to such an extent that they proposed the name moseleyum for the hoped-for element, in honor of Henry Moseley.[12] A plea to name the element quickly was also made by a Professor Hamer of the University of Pittsburgh in *Science Magazine*, claiming that this might avoid any repetition of the priority disputes that followed the discovery of hafnium. The wish to call it moseleyum was also supported in a letter to *Nature* magazine. But it was not to be, and neither has any other element been named after Moseley to this day, something that many consider to be a serious omission. The trouble with Bosanquet and Keeley's claim was basically "the same old story," in that nobody could reproduce the X-ray lines they claimed to have detected.

The Claimed Discovery of Masurium

A German team consisting of Walter Noddack and Ida Tacke, who later became his wife, made an extensive claim for having found not just element 43 but also element 75 in an article published in 1925. This claim would continue to have repercussions right up to the twenty-first century, when attempts were made to rehabilitate the work of the Noddacks. In fact, they had succeeded in discovering only one of the two elements, namely element 75, which they named rhenium after the river Rhine. Their claim for element 43, which they called masurium and which entered the literature and appeared in periodic tables in some countries, has not been sustained for the simple reason that the discovery of this element had to await its artificial synthesis following the discovery of nuclear fission.

It is interesting to examine the article written by the Noddacks in order to glean something of their methodology. First of all, they make a number of statements about the relative occurrence of the elements in the central portion of the periodic table and recognize the fact that the missing two elements in the manganese group will not necessarily resemble manganese, or at least that they will be more similar to the elements that flank them in a horizontal direction in the periodic table. The authors then present a fairly detailed table (fig. 6.2), in which estimates of the abundances of many elements are given. The conclusion is once again that one should not expect to

Sc	Ti	V	Cr	Mn	Fe	Co	Ni	Cu	Zn	Ga	Ge	As
2.10^{-3}	3.10^{-5}	3.10^{-5}	7.10^{-2}	10^{-2}	3.10^{-6}	3.10^{-5}	10^{-7}	10^{-6}	10^{-9}	10^{-7}		
Y	Zr	Nb	Mo	**43**	Ru	Rh	Pd	Ag	Cd	In	Sn	Sb
1.10^{-6}	6.10^{-5}	10^{-7}	10^{-7}	$\approx 10^{-13}$	2.10^{-12}	10^{-11}	10^{-11}	10^{-9}	10^{-8}	10^{-9}	7.10^{-6}	7.10^{-8}
La	Hf	Ta	W	**75**	Os	Ir	Pt	Au	Hg	Tl	Pb	Bi
6.10^{-7}	6.10^{-6}	5.10^{-7}	5.10^{-7}	$\approx 10^{-12}$	2.10^{-11}	2.10^{-11}	10^{-9}	10^{-9}	10^{-9}		4.10^{-7}	10^{-9}
	Th		U									
	7.10^{-8}		7.10^{-8}									

FIGURE 6.2 Relative abundance of the elements in the earth's crust as reported by W. Noddack and I. Tacke in 1925, including estimates for elements 43 and 75. Noddack, W. Tacke, I. Berg, O. Die Ekamangane, *Naturwissenschaften* 13 (26): 567–574, 1925.

find eka-manganese and dwi-manganese in the ores of manganese, whereas they are more likely to be found in ores of molybdenum and ruthenium (for element 43) and tungsten and osmium (for element 75).

The Noddacks caused a great deal of resentment due to the name they chose for what they believed was element 43, namely masurium. In a book on the elements written by the British chemist J. Newton Friend, the author wrote:

> The choice for element 43 was a stupid psychological blunder, which no civilized scientist should make. It commemorates the crushing defeat inflicted on the Russians by the Germans in the Masurian district during the Great War of 1914–1918, and thus tends to perpetuate racial hatred in a realm where such should be forgotten in a noble attempt to serve mankind.[13]

If this were not enough, the Noddacks remained in their academic positions during the Nazi regime in World War II, all of which contributed to there being great suspicion and in some cases all-out hostility toward their scientific claims to having discovered masurium.

Van Assche's Attempt to Rehabilitate the Noddack-Berg Claim for Element 43

The Belgian physicist Pieter Van Assche published an article in 1988 in which he claimed to rehabilitate the work of the Noddacks on the element they called masurium.[14] Van Assche begins by pointing out that Ida Noddack had correctly predicted the possibility of nuclear fission some five years before it was actually discovered. The author then moves on to analyze three kinds of arguments.

The first argument is referred to by Van Assche as a credibility test. Van Assche accepts that if any element 43 was present in the ores that the Noddacks and Berg examined, then it must have been formed by the spontaneous fission of uranium. Van Assche, therefore, tries to establish a correlation between the ores that the Noddacks experimented upon and their uranium content according to present estimates. Van Assche argues that this approach lends good support to the original Noddacks-Berg claim since they reported that they had found element 43 in the columbite, gadolinite, fergussonite, and sperrylite—all of which do indeed contain uranium, with the possible exception of sperrylite, for which current analyses are inconclusive. On the other hand, Van Assche points out, the Noddacks and Berg claimed that they did not find element 43 in platinore, tantalite, or wolframite and indeed these ores do not contain uranium according to current knowledge. As Van Assche writes,

> From table 3 we deduce that Noddack, Tacke and Berg observed element 43 in ores that nearly all are quoted to contain uranium; absence of element 43 goes together with absence of uranium. It is hard to imagine a better credibility test.[15]

The second of Van Assche's arguments concerns the actual X-ray spectrum in which the Noddacks and Berg claimed to have observed evidence for element 43. First of all, Van Assche mentions that the Noddacks calculated their experiment to have a detection limit of 10^{-9}, whereas he recalculates their detection limit to have been far lower—more like 10^{-12}. Van Assche also republishes an image of the X-ray spectrum in which the Noddacks claimed to have detected some lines that could be attributed to element 43. The author asserts that the three reported lines labeled $K_{\alpha 1}$, $K_{\alpha 2}$, and $K_{\beta 1}$ are extremely close to the values expected according to calculations, as are their relative intensities of 100:53:26, respectively.

Next, Van Assche gives an estimate of the abundance of element 43 expected on the basis of its formation from the spontaneous fission of uranium. By using the value of the half-life of ^{99}Tc (2.1 x 10^5 years), the half-life of ^{238}U (6 x 10^{15} years), and assuming that the Noddacks' sample of columbite contained about 5 percent of uranium, Van Assche arrives at a figure of about 10^{-13} for the abundance of element 43. This value is just one order of magnitude lower than the corrected detection limit that Van Assche has calculated, namely 10^{-12} as mentioned above. Van Assche's conclusion is that the Noddack experiments were easily capable of detecting element 43, since only one order of magnitude would seem to separate the detection limit of their X-ray experiment from the natural abundance of the element produced by the fission of uranium. Again, to quote Van Assche,

> As a conclusion, we state that at least one of the ores, discussed in detail by Noddack, Tacke and Berg, the searched for element 43, has a relative atomic abundance in the order of 10^{-13}.

Van Assche concludes his article by affirming the priority of the Noddacks and Berg, and even suggests that the element should once again be referred to as masurium, the name proposed by the Noddacks.

> These authors [Noddack, Tacke, Berg] eventually discovered the last missing stable element (75) and more than 14 years before the discovery of fission itself, the first fission product ($^{99}43$). It is very unfortunate that the well-documented discovery of element 43 was and still is being ignored. We hope that our analysis of Noddack, Tacke and Berg's results will restore the confidence they deserve; reintroducing the original name masurium (Ma) might be a tribute to the memory of these three exceptional scientists.

This article did not remain unchallenged for very long. In the very next year, Günter Herrmann of the University of Mainz in Germany (no nationalism here), refuted all of Van Assche's claims in great detail. Herrmann points out that in spite of their earlier claims, the Noddacks did not maintain that they had obtained element 43 for very long. For example, he mentions that, in her "last word on masurium," Ida Noddack admitted that it had proved impossible to extract the element in pure form from minerals. Turning to the specific claims made by Van Assche, Herrmann deals with each point individually.

On the "credibility argument" concerning the absence or not of uranium in the ores examined by the Noddacks, Herrmann points out that Van Assche has failed to comment on a further thirteen ores that were examined by the Noddacks and in which no element 43 was reported. Herrmann also mentions that the Noddacks reported a uranium content of several percent not only in columbite but also in tantalite, in which element 43 was not claimed to occur. Herrmann points out that Tacke later listed a total of twenty-seven minerals that he and the Noddacks had examined, whereas Van Assche is basing his credibility argument on just six or seven ores.

On the question of the X-ray image Herrmann points to some discrepancies, such as the fact that the distances between the assigned three lines are equal in the X-ray image, whereas in a grain size graph published alongside the spectrum the distances increase by factors of 1:2:3. Herrmann also reminds readers of the fact that the story surrounding the X-ray plate is somewhat mysterious. For example, when Segrè asked to see the plate in 1937 he was told that it had been destroyed. Herrmann concludes by saying that he sees

> no argument for a thorough revision of the history of element 43 in favour of an early discovery by W. Noddack, I. Tacke and O. Berg.

However, it appears that Van Assche did not accept Herrmann's refutation. A few years later he approached the US geochemist John Armstrong, who was working at the National Institute of Standards and Technology (NIST) in Gaithersburg, Maryland, and asked for his help in analyzing the alleged X-ray spectrum of element 43. Not only did Armstrong support Van Assche, but he undertook the task of simulating the X-ray spectrum of element 43 according to Van Assche's calculations on the uranium content of the Noddack samples from 1925.

The outcome of this collaboration was two published articles, although neither appeared in the primary research literature. The first was an anonymously authored note of less than one page in length in the November–December 1999 issue of the NIST journal.[16] The note begins by repeating the fact that the claimed discovery of element 43 by the Noddacks and Berg is invariably derided and dismissed. It then reads:

> To test if the claims of Noddack et al. are plausible, a NIST scientist utilized the NIST x-ray database and spectral analyzer program DTSA to simulate the 1925 data. The experimental configuration was deduced from the original paper and simulations were made for a range of compositions for the residue suggested by Van Assche. The relative intensity of various x-ray lines and the peak-to-background ratios were determined and compared to the original spectrum.

The conclusion of the note is:

> The NIST simulation provides compelling support for the 1925 spectral claims and demonstrates the forensic capabilities enabled by recent advances in measurement science.

The other published item[17] was an entry in the otherwise excellent compilation of articles commissioned by the US magazine *Chemical & Engineering News* in order to celebrate the 80th anniversary of the publication.[18] In this article, John Armstrong writes:

> It wasn't until 1998 that I took a real look at element 43 . . . One day, an exuberant Belgian physicist, Pieter Van Assche, came into my office to ask my interpretation of an X-ray spectrum. The spectrum was from a 1925 article by Ida Noddack, Walter Noddack and Otto Berg, who claimed to have discovered element 43 (which they called "masurium") in samples from uranium-rich ores.

He continues:

> Using first-principles X-ray-emission spectral-generation algorithms developed at NIST, I simulated the X-ray spectra that would be expected for Van Assche's initial estimates of the Noddacks' residue compositions. The first results were surprisingly close to their published spectrum! Over the next couple of years, we refined our reconstruction of their analytical methods and performed more sophisticated simulations. The agreement between simulated and reported spectra improved further. Our calculation of the amount of element 43 required to produce their spectrum is very similar to the direct measurements of natural technetium abundance in uranium ore published in 1999 by Dave Curtis and colleagues at Los Alamos. We can find no other plausible explanation for the Noddacks' data than that they did indeed detect fission "masurium."
>
> The Noddacks were clearly among the finest analytical geochemists of their time. Their search for the "missing" elements below manganese in the periodic table was part of a larger effort to accurately determine the abundance of the chemical elements in the earth and meteorites—data that provided a foundation for the science of geochemistry. Their work complemented rather than detracted from that of Perrier and Segrè.

Nothing was heard again of this debate until an article appeared in 2005 in the *Journal of Chemical Education*.[19] Here, an Italian author, Roberto Zingales, gave a history of the discovery of element 43 and concluded his account by mentioning that,

> As the ores studied by the Noddacks may have contained as much as 10% uranium, in recent years, John T. Armstrong of the NIST, tried to duplicate their experiments. Using a sophisticated X-ray database and spectral analysis software, Armstrong found that the spectral lines attributed to masurium appear consistent with

> element 43 *(39)* and that Noddacks' instruments were sufficiently sensitive to detect less than a billionth of a gram of this element *(13)*. In conclusion, it took about 75 years to remove the doubt cast on the Noddacks' results and to finally agree on the reliability of their scientific results. It is an irony of fate that nuclear fission, whose existence was firstly suggested by Ida Tacke Noddack, was used to vindicate them.

A response was not late in coming. Fathi Habashi, a professor of metallurgy from Laval University in Quebec, Canada, sent a letter to the editor pointing out that the claim by Van Assche and Armstrong "cannot stand up to the well-documented assertion of the well-established physicist Paul Kuroda (1917–2001) in his paper 'A Note on the Discovery of Technetium.' "[20] He was referring to Kuroda's work on the amount of technetium present.[21]

The same issue of *Journal of Chemical Education* of 2006 also carried a retraction by Roberto Zingales in which he mentions having received a letter from Herrmann and the fact that he now realizes that the claims of Van Assche and Armstrong on behalf of the Noddacks and Berg cannot be sustained.[22]

Finally, the "Real" Element 43 is Obtained

Following all the failed attempts at isolating element 43, including that of the Noddacks and Berg, the element was finally and genuinely obtained in 1937. It was a product of a synthetic process rather than a case of isolation from naturally occurring minerals. It was also the first time that an element was discovered following its artificial synthesis.

Although very small amounts of element 43 were subsequently found to exist naturally, the first discovery of the element was carried out using a plate of molybdenum metal that had been irradiated with deuterons and neutrons. Unlike the synthetic elements that are created these days, the discovery of element 43 was somewhat serendipitous in that it did not follow a deliberate attempt to create the element. Instead, the researchers happened to find it while analyzing a sample that had undergone irradiation such as to bring about the transmutation of one element, molybdenum, number 42, to the following element, number 43. This is not to say that the physicists and chemists concerned were completely oblivious to the possibility of the presence of element 43. Rather, they quickly suspected that it might be present from knowledge of radioactive processes and set about trying to detect the possible existence of element 43.

The task fell to Italian physicist Emilio Segrè, who had spent a postdoctoral fellowship at the Berkeley lab of Ernest Lawrence, the creator of the cyclotron machine. Segrè then returned to his permanent job in Palermo, Sicily, and was sent an irradiated sample of molybdenum by a fellow countryman who was still in California. Together with the chemist Carlo Perrier, Segrè undertook the task of eliminating all other possible elements, given that many others were also produced in the original irradiation. As Segrè wrote,

> In the search for the four missing elements lower than uranium, we have now been able to prepare element 43 and study its properties in detail. The substance was produced in the Berkeley cyclotron which thus remains for the time being the only "mine" of the element.[23]

Segrè and Perrier succeeded in isolating the isotopes technetium-95 and technetium-97. In 1940, Segrè and Wu also found element 43 to be a product of the fission of uranium. The University of Palermo officials wanted them to name their discovery panormium, after the Latin name Panormus for Palermo. But element 43 was named after the Greek word τεχνητός, in 1947, meaning "artificial," since it was the first element to be artificially produced.

In doing this, the authors were following the suggestion by Paneth, who wrote in the very same issue of *Nature* magazine. It is worth pausing to consider Paneth's paper since it reveals a hidden aspect of the story of a number of our seven elements (fig. 6.3). The simple fact is that for many years the artificially synthesized elements were not considered as true elements by all chemists. In many cases, the newly discovered elements remained nameless and did not appear on the periodic table. There was a general feeling in some circles that synthetic elements were somehow quite different from naturally occurring ones. The initiative to change this situation was taken by Fritz Paneth. Paneth begins by alluding to one of his own lectures.

> Almost five years ago, in a lecture to the Institute of Chemistry in London, the success that radioactive methods had achieved in the task of completing the Periodic System was described. In the table given, the place of element 87 was filled by the symbol of a newly discovered branch product[24] in the actinium series; but in the places of the elements 43, 61, 85 and 93 no symbols were inserted, although, as explained in some detail, atoms of all of these four elements had been artificially produced.[25]

Paneth then suggests that whereas the denial of "full citizenship" to artificial elements may have been justified in his 1942 lecture, this was no longer the case in 1947. In 1942, the elements that had been produced artificially were all unstable, had only been produced in invisible amounts, and were not present on the earth. In the intervening five years, several pounds of one particular element, plutonium, number 94, had been stockpiled. Secondly, its half-life of some 200,000 years ensured that plutonium would remain on Earth for many years to come! Finally, the developing technology made it clear that many more elements, with even higher atomic numbers, could soon be produced. Paneth then writes:

Periodic Classification of the Chemical Elements

Period	Group																	
	1	2	3	4	5	6	7	8	9	10	11	12	13	14	15	16	17	18
I																	1 H 1.0080	2 He 4.003
II	3 Li 6.940	4 Be 9.02											5 B 10.82	6 C 12.010	7 N 14.008	8 O 16.0000	9 F 19.00	10 Ne 20.183
III	11 Na 22.997	12 Mg 24.32											13 Al 26.97	14 Si 28.06	15 P 30.98	16 S 32.06	17 CI 35.457	18 A 39.944
IV	19 K 39.096	20 Ca 40.08	21 Sc 45.10	22 Ti 47.90	23 V 50.95	24 Cr 52.01	25 Mn 54.93	26 Fe 55.85	27 Co 58.94	28 Ni 58.69	29 Cu 63.57	30 Zn 65.38	31 Ga 69.72	32 Ge 72.60	33 As 74.91	34 Se 78.96	35 Br 79.916	36 Kr 83.7
V	37 Rb 85.48	38 Sr 87.63	39 Y 88.92	40 Zr 91.22	41 Nb 92.91	42 Mo 95.95	43—	44 Ru 107.7	45 Rh 102.91	46 Pd 106.7	47 Ag 107.880	48 Cd 112.41	49 In 114.76	50 Sn 118.70	51 Sb 121.76	52 Te 127.61	53 I 126.92	54 X 131.3
VI	55 Cs 132.91	56 Ba 137.36	57-71 Rare Earths†	72 Hf 178.6	73 Ta 180.88	74 W 183.92	75 Re 186.31	76 Os 190.2	77 Ir 193.1	78 Pt 192.23	79 Au 197.2	80 Hg 200.61	81 TI 204.39	82 Pb 207.21	83 Bi 209.00	84 Po 210	85—	86 Rn 222
VII	87 AcK 223	88 Ra 226.05	89 Ac 227	90 Th 232.12	91 Pa 231	92 U 238.07	93—											

† Rare Earths

VI 57-71	57 La 138.92	58 Ce 140.13	59 Pr 140.92	60 Nd 144.27	61—	62 Sm 150.43	63 Eu 152.0	64 Gd 156.9	65 Tb 159.2	66 Dy 162.46	67 Ho 164.94	68 Er 167.2	69 Tu 169.4	70 Yb 173.04	71 Lu 174.99

FIGURE 6.3 From Paneth's article of 1942. F. A. Paneth, Radioactivity and the Completion of the Periodic System, *Nature, 149*, 565–568, 1942. Reprinted with permission of Nature Publishing Group.

> In these circumstances, there seems to be no doubt that the time has come for the systematizing chemist no longer to discriminate between natural and artificial elements, but to pay equal attention to the study of both and, in the tables of the Periodic System, to insert everywhere the appropriate symbols.

Paneth suggests a set of rules for assigning names and symbols for elements 43, 61, 85, 87, 93, 94, 95, and 96.

1. The right to name an element should go to the first to give definitive proof of the existence of one of its isotopes.
2. In deciding the priority of the discovery, there should be no discrimination between naturally occurring and artificially produced isotopes.
3. If a claim to such a discovery has been accepted in the past, but refuted in later research, the name given should be deleted and replaced by one chosen by the real discoverer.

He then comments on the fact that the need for the last rule arises because there have already been at least two cases of its violation.

> The names "masurium" and "illinium" are so firmly rooted in textbooks and tables that recent work on artificial isotopes of the elements 43 and 61 is sometimes referred to as the production of species of masurium and illinium, while the artificial isotopes actually were the first representatives of the hitherto missing chemical elements.

At this point, Paneth launches into something of a tirade against the Noddacks in particular. He laments the failure of claimants to withdraw their statements even though many years of intensive efforts had failed to support their claims. He adds that Walter Noddack even went so far as to complain to the convener of a chemical meeting for not having invited him to speak on this element, as the geographical location of the meeting would have been an opportunity to expound on the chemistry of masurium.[26] He further mentions that during the war, when Noddack was appointed professor of inorganic chemistry by the occupying powers in Strasbourg, the symbol Ma continued to occupy the space for element 43 on the periodic table of the chemistry lecture hall.[27] After making it clear that he believes Segrè and Perrier are the discoverers of element 43, and similarly emphasizing the true discoverers of elements 61, 85, and 93, Paneth concludes his article with the following invitation to his colleagues:

> So far no names for elements 43, 61 and 85 have officially been put forward by their discoverers, Perrier and Segrè, Coryell and his group, and Corson, MacKenzie and Segrè respectively. Every chemist concerned with the task of teaching systematic inorganic chemistry and of keeping his table of the Periodic System up to date will be grateful if they will publish soon the names which they consider suitable.

The editor of *Nature* magazine appears to have acted immediately since in the very same issue two of these three teams published letters in which they officially unveiled their proposed names and referred to Paneth's proposal in doing so.[28]

Segrè later returned to Berkeley and met Glenn T. Seaborg, where in 1938 they isolated the metastable isotope technetium-99-m—now used in millions of medical diagnostic procedures throughout the world.[29] To complete this section, here is what Segrè said in a later oral interview about the masurium episode.[30]

> Well, you see, there was a complicated story. Years after—In '37, '38 and '39—I became convinced, and quite properly, that the Noddacks were a mixed type of people. They had done excellent work on rhenium but they had been plain dishonest—there's no other way of saying it—on technetium 43, or what they called

masurium. You see, at first they thought—and it could happen to anybody—they thought they had discovered it. But then I visited them, and I saw them and I talked to them, and after having seen them and talked with them, by '37, I was convinced that they were dishonest.

I even have a letter from Hevesy saying that he also has the same conviction. Well, they were not really quite dishonest; they were in this situation: Their work on rhenium was good—there was no doubt about it. It was excellent, and they had discovered rhenium. The work on masurium just wasn't right. And now we know there is no stable masurium. But they had a hope. By that time (1937), they must have felt that it was not right, but they had the hope that maybe after all, element 43 might turn up. And then they said, "All right, if we say nothing retracting our earlier work, we can always say that we have discovered it." And so they never took it back. They died without ever taking it back, always saying that they had seen masurium.

Oklo Phenomenon

In 1939, at the time when the study of nuclear reactions was still in its infancy, an article was published by S. Flügge in which this author speculated about the possibility of a self-sustaining chain reaction involving the element uranium that might have taken place under natural conditions in the past.[31] Flügge also suggested that such an event might have taken place in the large uranium deposit in St. Joachimstahl, Bohemia, or alternatively in the carnotite deposits in the US state of Colorado.

The first man-made sustained nuclear reactor was built at the University of Chicago in 1942 by the Italian-born physicist Enrico Fermi. The success of Fermi's achievement, and the scientific and engineering demands of this work, was interpreted by many physicists to mean that any such natural process would not have been possible.

Nevertheless, very strong evidence of such a natural process was discovered in the 1970s, by the French Atomic Energy Establishment working in the African republic of Gabon. Even more remarkably perhaps, the age of such a pre-Fermi reactor, as it became known, was predicted with amazing accuracy in 1965 by Japanese nuclear chemist Paul Kuroda, working at the University of Arkansas.

By making a few basic assumptions, Kuroda reached the conclusion that the natural reactor would have been active about 2×10^9 years ago. Following the French discovery of the Oklo reactor in the Gabon, the calculated age of the event was placed at precisely 2×10^9 years ago, in agreement with Kuroda's prediction. The reaction of the physics community was one of complete surprise. For example, Clyde Cowan, the codiscoverer of the neutrino, wrote,

> The announcement of the Oklo reactor was received by American nuclear scientists with skepticism. Some of the world's best physicists had constructed the Stagg Field reactor with careful attention to mechanical detail, to the purity of the materials and the geometry of the assembly. Could nature have achieved the same result so casually? We now know that the answer is yes.[32]

The Oklo reactor was discovered in June of 1972 as a result of a significant anomaly in the ratio of the two main isotopes of uranium, namely ^{235}U and ^{238}U. Whereas the normal fraction of ^{235}U in naturally occurring uranium is 0.7202 percent +/- 0.0010 percent, the minerals from Oklo were found to contain 0.440 percent of the same isotope. The conclusion arrived at by the French nuclear scientists was that a self-sustaining nuclear chain reaction had taken place at the time that had been predicted by Kuroda. Estimates of how long the natural nuclear reactor had existed varied between 600,000 and 1.5 million years. In addition, it was deduced that the presence of oxygen in the atmosphere had participated in the fractionation processes occurring within the uranium minerals. This also coincided with the independent estimates that the Earth's atmosphere underwent a dramatic enrichment in oxygen 2 x 10^9 years ago as a result of the generation of new living organisms capable of the process of photosynthesis.

Further analysis of the minerals in the Oklo reactor showed that a number of elements that had previously been thought to be absent from the Earth did in fact occur as a result of such unusual nuclear phenomena. These elements included technetium, promethium, and even element 93, or plutonium.[33]

Technetium in Space and Technetium Applications

Turning to the skies, technetium was detected in some so-called red giant stars as long ago as 1952, but not in our own sun, a fact that has had a significant role in confirming the view that the Sun is a relatively young star. In addition, because technetium isotopes have short half-lives compared with the age of red giants, this finding suggested that the element was being produced within the stars, thus supporting theories of stellar nucleosynthesis for elements of intermediate mass.

In spite of its exotic heritage, technetium is now widely used in medicine as a diagnostic tool. Radioactive molybdenum-99 is allowed to decay to form technetium-99m—meaning ^{99}Tc in an excited nuclear state. This metastable isotope then drops to the ground state with the loss of a gamma particle, which can be recorded in radiodiagnostic procedures for the detection of tumors, among other things.

The usefulness of ^{99}Tc lies in a number of specific properties that it has. The radioactive decay of the excited form has a half-life of six hours; that is long enough to be injected into a patient before it decays, but still sufficiently short for its emission intensity to be measurable at low concentrations. Furthermore, the short half-life means that the patient need only be exposed to radiation for a brief period of time. The aqueous chemistry of technetium is also critical. The pertechnate ion TcO_4^-, which is the form in which the isotope is administered, is both soluble and stable over the physiological range of interest, unlike, for example, the permanganate ion (MnO_4^-).

Finally, one of the best ways to protect steel from corrosion when it is in contact with water, even at high temperatures and pressures, is to use a coating of $KTcO_4$. It is unfortunate that technetium is radioactive, otherwise this method could be used in situations other than with steel in carefully sealed containers.

Medical Imaging

The main application of the element technetium lies in medical imaging. This involves one particular isotope of the element and moreover a metastable nuclear isomer of the same isotope, namely Tc-99m. The nuclear isomer was discovered by Segrè and Seaborg in 1938 but its widespread applications had to wait until far more recently (fig. 6.4).

FIGURE 6.4 McMillan, Segrè and Seaborg. By permission from Emilio Segrè Collection at the Institute of Physics.

Medical imaging using Tc-99m accounts for about 85 percent of all medical imaging diagnostic tests these days, with more than 50 million procedures conducted each year worldwide.

The nuclear isomer is radioactive, with a half-life of about six hours, a feature that makes it very suitable for use with human subjects. In a 24-hour period, something like 94 percent of the isomer decays from the body. It is also suitable because in the course of decaying from a metastable to a stable state, the isomer emits gamma rays with an energy of 140 keV, which happens to be comparable with the radiation output from a commercial diagnostic X-ray machine. The more stable isomer of Tc-99 eventually decays to ruthenium-99 although the half-life of this process is a huge 211,000 years and is accompanied by harmless electron emission as in all β decay processes.

Technetium Generators

These days all major hospitals possess some technetium generators containing molybdenum-99, which is the parent nuclide that produces technetium-99m. The radioactive molybdenum in turn comes from the decay of highly enriched uranium. Technetium generators have been used for many purposes, including imaging of the brain, myocardium, thyroid, lungs, liver, gallbladder, kidneys, skeleton, blood, and tumors.

The molybdenum-99 that is the fuel in technetium generators has a half-life of approximately 2.75 days with the result that these generators can be shipped anywhere in the world and will still produce reasonable amounts of T-99m after a period of a week. When technetium is produced in a generator, it is released in the form of the pertechnate ion, or TcO_4^-, in which technetium displays an oxidation state of +7 and is not suitable for medical applications. Before it can be useful, it needs to be reduced chemically to a lower oxidation state and the technetium must be coordinated with a ligand so that it will have an affinity with any particular organ of interest.

Chemistry of Tc

The fact that macroscopic quantities of technetium are available and that ^{99}Tc only emits harmless electrons rather than gamma rays has led to its chemistry being thoroughly studied, with a view to finding compounds that will bind with every conceivable part of the human body.

Before discussing these medically useful compounds, it is worth pausing to consider some simple compounds of technetium. Just as Mendeleev predicted in 1881, the element forms colored oxides with the formulas of TcO_2, TcO_3, and Tc_2O_7. As many as nine oxidation states have been identified in compounds of technetium, ranging from –1, through 0 to +7.[34]

Coordination Compounds of Tc

The chemical and biological properties of technetium compounds are largely governed by the ligands that they are bonded to in coordination compounds. The first set of such compounds developed were useful for imaging the excretory organs as well as bones. These "first generation" technetium compounds include examples such as ^{99m}Tc DTPA, where the ligand DTPA stands for diethylenetriamine pentaacetic acid (fig. 6.5).

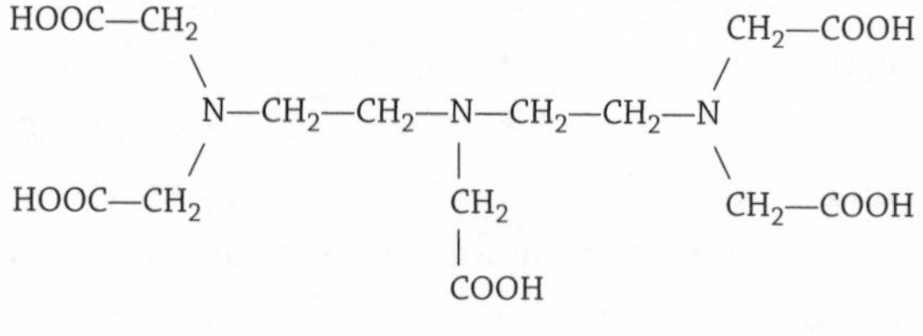

```
HOOC—CH2                                    CH2—COOH
        \                                  /
         N—CH2—CH2—N—CH2—CH2—N
        /              |                   \
HOOC—CH2              CH2                   CH2—COOH
                       |
                      COOH
```

FIGURE 6.5 DTPA or diethylenetriamine pentaacetic acid ligand.

The second generation of technetium compounds was developed in order to medically image the heart muscle (myocardium) as well as the brain. One such compound is ^{99m}Tc HMPAO (hexamethylpropylene amine oxymatooxo technetium(V) with trade name of Ceretec®).

Chapter 7

Element 87—Francium

H																	He
Li	Be											B	C	N	O	F	Ne
Na	Mg											Al	Si	P	S	Cl	Ar
K	Ca	Sc	Ti	V	Cr	Mn	Fe	Co	Ni	Cu	Zn	Ga	Ge	As	Se	Br	Kr
Rb	Sr	Y	Zr	Nb	Mo	Tc	Ru	Rh	Pd	Ag	Cd	In	Sn	Sb	Te	I	Xe
Cs	Ba	Lu	Hf	Ta	W	Re	Os	Ir	Pt	Au	Hg	Tl	Pb	Bi	Po	At	Rn
Fr	Ra	Lr	Rf	Db	Sg	Bh	Hs	Mt	Ds	Rg	Cn		Fl		Lv		

La	Ce	Pr	Nd	Pm	Sm	Eu	Gd	Tb	Dy	Ho	Er	Tm	Yb
Ac	Th	Pa	U	Np	Pu	Am	Cm	Bk	Cf	Es	Fm	Md	No

FIGURE 7.1 Showing position of element 87, or as eventually named, francium.

Quick Overview

One of the most remarkable things about element 87 is the number of times that people claimed to have discovered it after it was predicted by Mendeleev in 1871 and given the provisional name of eka-caesium (fig. 7.1). It was recognized early on that the periodic table more or less fizzles out after element 83, or bismuth. All subsequent elements are radioactive and therefore unstable, with a few exceptions like uranium and thorium. But this fact did not deter a number of scientists from searching for element 87 among natural sources and in many cases from claiming to have isolated it. For example, Druce and Loring in England thought they had identified the element by using the classic method developed by Moseley for measuring the K_α and K_β lines of any element's X-ray spectrum. But it was not to be.

In the 1930s, it was the turn of Professor Fred Allison from the Alabama Polytechnic Institute (now Auburn University). Allison developed what he called a magneto-optical method for detecting elements and compounds based on a supposed time lag in the development of the Faraday effect, whereby the application of a magnetic field causes a beam of polarized light passing through a liquid solution to be rotated. Allison mistakenly thought that every element gave a particular time lag, which he claimed was observed with the naked eye, and that this effect could be used to identify each substance. He boldly claimed in a number of journal articles, and even a special feature in *Time Magazine*, that he had observed elements 87 and also 85, both of which were still missing at the time.

Literally hundreds of papers were published on this effect, including a number of studies arguing that it was spurious. But these days the Allison effect is often featured in accounts of pathological science, alongside the claims for N-rays and cold fusion.[1]

The next major claim came from Paris and was supported by Jean Perrin, the physicist who is best known for having provided confirmation for Einstein's theory of Brownian motion and for thus providing supporting evidence for the real existence of atoms. A Romanian physicist working with Perrin, Horia Hulubei, claimed to have used highly accurate X-ray measurements and to have recorded several spectral lines with precisely the frequencies expected of element 87, which he promptly named moldavium. But alas, these lines too turned out to be spurious.

The eventual discovery of element 87 was made by a remarkable Frenchwoman, Marguerite Perey, who began life as a laboratory assistant to none other than Marie Curie in Paris. Perey became skillful in purifying and manipulating radioactive substances and was asked to examine the radioactivity of actinium, which is element number 89 in the periodic table. By carefully excluding all daughter isotopes, she was the first to observe the α and β radiation produced by actinium itself rather than its radioactive daughter isotopes and thereby discovered a weak but significant branch in one of the three main radioactive decay series. Her analysis of the data revealed a new element with a half-life of twenty-one minutes. When she was later asked to name the element, she chose francium to honor the country of her birth.[2] It was also an appropriate choice in marking the continuing contribution made by French scientists to the study of radioactivity. After all, the phenomenon was first discovered by Becquerel, the Curies isolated polonium and radium, and Debierne isolated actinium, all within a few years of each other and all in France.

As it turns out, francium was the last natural element to be discovered. Estimates of the abundance of francium suggest that there is only about 30g in the whole of the earth's crust. It is one of a very few elements that has no commercial applications, mainly because its longest-lived isotope has a half-life of just twenty-one minutes. Nevertheless, the fact that the francium atom has the largest diameter of any element, at an outstanding 2.7 Ångstroms and the fact that it has just one outer-shell electron, have made it the object of considerable attention among researchers wanting to probe the finer details of current theories of atomic physics. In 2002, a group in the United States succeeded in trapping 300,000 atoms of francium and in performing several key experiments of this kind.[3]

The Search for Element 87 in More Detail

Reports on the discovery of element 87 first appeared in 1925 in the November 6 issue of *Chemical News*. F. H. Loring and J. G. F. Druce referred to an earlier paper authored by Druce in which the aim had been to investigate the possible presence of element 43 and 75, and even 93, in various minerals. As the authors state:

> on the isolation of crude rhenium oxide, it was noticed that a strange line appeared on one of the photographic films of the X-ray spectrum. . . . on investigation we found that this line of wave-length 1.032Å units, fell exactly between the theoretical $L_{\alpha 1}$ and $L_{\alpha 2}$ lines of element of atomic number 87.

How the authors did this is rather interesting. They used the known wavelength difference between the $L_{\alpha 1}$ and $L_{\alpha 2}$ of copper in order to estimate where the $L_{\alpha 1}$ and $L_{\alpha 2}$ lines of element 87 should fall. Doing so gave them an estimate of 1.0275Å for the wavelength of the $L_{\alpha 1}$ of element 87, which is remarkably close to the theoretical value of 1.027Å that is obtained by extrapolation from the values for neighboring elements using Moseley's method and as shown in fig. 7.2.

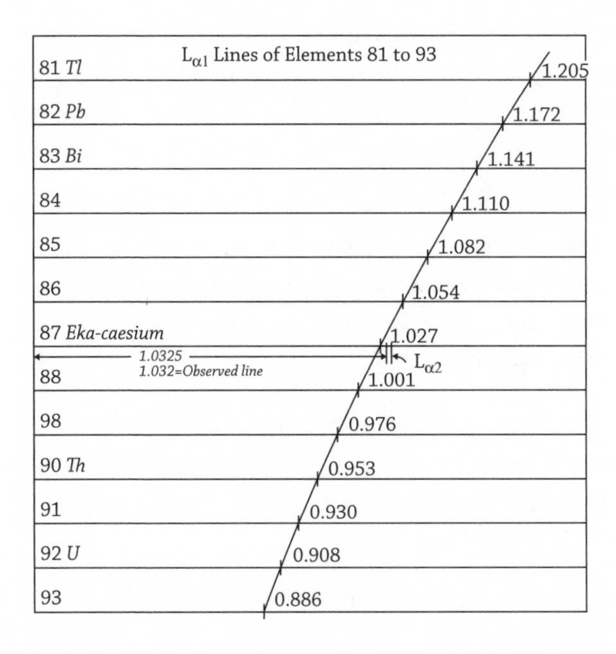

FIGURE 7.2 Loring and Druce diagram showing expected wavelengths of $L_{\alpha 1}$ lines for elements 81 to 93. From F. H. Loring, J. G. F. Druce, Eka-Caesium, *Chemical News, 31,* 289–289, 1925.

It is not surprising that Loring and Druce believed that they had been successful and could reasonably claim that[4]

> We have, for the present, designated the element *eka-caesium* in accordance with the nomenclature adopted by Mendeleeff. The existence of the highest and apparently the last homologue of the alkali-metal group (I), is thus indicated. [5]

Nevertheless, they also reported a failure to observe any $L_{\beta 1}$ lines for the element and claimed,

> The evidence that the line is an α line of element of atomic number 87 is very strong, for in the immediate region of this line there are no other lines which would come out with same intensity in all probability.[6]

Just a week later, on November 13, a further note appeared, this time entitled "Eka-caesium and eka-iodine." While reporting that they were still unable to obtain the $L_{\beta 1}$ line for element 87, they now announced a further success, namely the discovery of another of our missing seven elements, element 85.

> In addition thereto, on the same film, lines of wave-lengths 1.086 and 0.895 were obtained, though very feint. These would seem to correspond with the characteristic L_{α} and L_{β} radiations of element 85, that is eka-iodine.

Once again their optimism seems very reasonable since the curve predicts 1.082Å for the $L_{\alpha 1}$ line of eka-iodine. Using another sample, consisting of pyrolusite, Loring and Druce also reported a new line of wavelength 1.040Å,

> which allowing for width would include the $L_{\alpha 1}$ and $L_{\alpha 2}$ radiations of eka-caesium.

This last claim seems a little surprising since according to the curve of their previous article (fig. 7.2), a wavelength of 1.040Å would lie closer to the value expected for element 86, or radon. If anything, this finding seems to spoil their previous more accurate claim for the X-ray line at 1.032Å.

In any case, although the eka-iodine (element 85) claim appeared conclusive, the authors had the good sense to show considerable caution when writing:

> With regard to the eka-iodine lines, which seemed to offer conclusive evidence, owing to the characteristic L_α and L_β lines appearing, we feel compelled nevertheless, to regard these lines as uncertain indications of this new element being present in the sample; but we will say with confidence that the research we have started on the missing elements in, or in connection with, Group VII., is promising, and the work is being continued.[7]

Referring to both elements 85 and 87 they concluded with:

> We fully realize that much more work is necessary to establish the two elements, here mentioned, on a firm basis.[8]

This set of reports is a good example of some reasonable claims being made on the basis of Moseley's method and showing an appropriate degree of caution. Such people who claimed the discovery of elements that eventually turned out to be spurious were by no means as naïve as one might be tempted to believe from a modern perspective.

Allison and Alabamium

Quote from *Time Magazine*:

> With an air of "There, that will convince them," Professor Fred Allison of Alabama Polytechnic Institute last week slapped on his

> desk a fresh copy of the American Chemical Society's Journal. "Them," referred to everyone who doubted that Professor Allison had discovered Element No. 87, or eka-cesium, in 1930 and Element No. 85, or eka-iodine last April by means of his new magneto-optical machine. "Them" referred particularly to Professor Jacob Papish of Cornell, who last autumn recognized eka-cesium with the x-ray spectrograph. With an x-ray spectrograph Professor B. Smith Hopkins of the University of Illinois discovered the third last unknown element, No. 61, of the Periodic Table, which he named illinium.[9]

This passage would seem to indicate the great confidence with which Professor Fred Allison, from Alabama Polytechnic Institute, regarded his experiments, wherein he claimed to have detected not just one, but two, of the missing seven elements of our story. Before discussing how he was incorrect in this and many other claims, and how this represents a textbook example of "pathological science," it is worth taking a moment to look into the science of Allison's magneto-optical method of chemical analysis.

Allison's method was based on the Faraday effect, which is not only a firmly accepted part of science but which presented the first evidence of the relationship between light and electromagnetic radiation. In 1845, Michael Faraday, the director of the Royal Institution in London, discovered that the plane of polarization of a light beam passing through a liquid undergoes a measurable rotation on the application of a strong magnetic field.

Many years later, Allison constructed an apparatus in which a light spark was generated and directed through crossed polarizers and then through two glass cells containing identical solutions of any given liquid. These cells were in turn surrounded by coils of wire and would serve to generate the magnetic field when a current passed through them. However, Allison was primarily interested in measuring the presence of a postulated time lag between the application of the magnetic field and the appearance of the Faraday rotation. He varied the time at which the magnetic field was applied by adjusting the distance through which the current had to travel in order to reach each of the glass cells. The delay in the Faraday effect would be observed by looking for a minimum in the light from the spark after the magnetic field had been switched on. The reasoning was that if the direction of polarization had been rotated, the light should show a certain depletion in brightness.[10]

Not only did Allison and his coworkers claim that they could visually detect such minima in different materials, but they succeeded in publishing literally dozens of articles in the leading scientific journals, including *Physical Review* and the *Journal of the American Chemical Society*. They claimed that the sensitivity of their apparatus was so great that they could detect time differences of the order of 10^{-10} of a second. They also claimed that in the case of ionic solutions such as potassium nitrate, sulfate, and chloride, the mere presence of the negative ion would modify the time lag, which was entirely due to the positive metal ion. Such claims met with much skepticism from other researchers but remarkably enough, other studies supporting Allison's claims were published by a number of laboratories. In all, hundreds of peer-reviewed articles were published on what became known as the Allison effect, until the work was debunked in a number of key studies and the journals decided to stop disseminating research on this topic. But this did not happen before Allison had gone even further by claiming to have discovered the long-sought element 85 and, later, element 87.

Refutations

In 1933 Wendell Latimer, who later became head of chemistry, at U.C. Berkeley, went to visit Allison in order to see if he could use the magneto-optical method to detect hydrogen atoms of atomic weight 3 that had begun to be discussed as a distinct possibility. Latimer even made a ten-dollar bet with G. N. Lewis, of chemical bond fame, the latter thinking that Latimer would not find anything by using Allison's method. On visiting Allison, Latimer duly learned the technique and claimed that he had indeed detected atoms of ^3H, or tritium as it subsequently became known. Again this article was successfully published in the *Physical Review* and Latimer went back to Berkeley, where he collected his ten dollars from G. N. Lewis.

In 1934, H. G. MacPherson at U.C. Berkeley published a paper in which he reported that Allison's minima were an illusion and ended the article by saying:[11]

> The author believes, as a result of these experiments, that such minima as he saw from time to time can be explained as due to physiological or psychological factors.

As months went by, Latimer realized[12] that he could not reproduce the alleged observation of ^3H atoms. Tritium was discovered later in 1934 by Rutherford, Oliphant, and Harteck.[13]

On a final note concerning Allison's claims, it is now estimated that the outer one mile of the earth's crust contains only 69 mg of element 85 and 24.5 g of element 87, as compared with as much as four thousand tons of polonium, which is itself considered to be an exceptionally rare element. Both elements 85 and 87 have extremely short-lived isotopes and in any case, Allison was looking for the elements in the wrong kinds of minerals.

Horia Hulubei

Horia Hulubei was born in Romania in 1896. During World War I he was a soldier in his country's army and managed to escape to France following the invasion of Romania by the Austro-German forces. This was to be the first of many visits to France that would bring him some scientific fame but not the discovery of element 87 that he so dearly wanted.

In this first trip to France he became a fighter pilot in the French Air Force, before returning to Romania to complete his education as a physicist. In 1927, he came back to France to join the physical chemistry laboratory of Jean Perrin at the Sorbonne in Paris.[14] Hulubei remained in Paris throughout the 1930s, collaborating with X-ray expert Yvette Cauchois. Their work together included concerted attempts to identify elements 85, 87, and even 93.

In 1936, Hulubei and Cauchois used a highly sensitive X-ray apparatus to report some weak lines believed by them to include an L_α doublet due to element 87 with wavelengths of 1032 and 1043 X.U. The following year Hulubei's claim was strongly criticized by F. H. Hirsh from Columbia University, who accused Hulubei of mistaking X-ray lines due to mercury for the supposed element 87.

Hirsh pointed out the existence of two mercury lines at 1030 and 1047 XU, plus the fact that Hulubei and Cauchois's wavelengths differed from them by less than their own published error margins. Hirsh then proceeded to describe some of his own experiments to search for the X-ray lines of element 87 in lepidolite in which he found no conclusive evidence whatsoever for such lines.

Just before the outbreak of World War II, Hulubei was back in Romania to take up an appointment as professor of physics at Iasi University. Because of the break in communications between Romania and the United States, Hulubei did not hear of Hirsh's criticism until the year 1943, but nevertheless undertook a response to the American's comments as late as 1947, a full ten years after the discovery in question.[15] In a letter to the editor of *Physical Review*, he begins by saying:

> The article by F.R. Hirsh "The Search for Element 87" came to my notice recently because of the interruption of postal services between America and Rumania during the war.

Hulubei turns the criticism right back at Hirsh by pointing out that the wavelengths of 1030 and 1047 for the lines of mercury, as cited by Hirsh, are in error according to Moseley's law. He also says,

> I suggest that his categorical objections to the inferences of other authors [for element 87] are based on very inexact estimates of the emission lines which are to be expected.

And when it comes to Hirsh's own experiments, Hulubei writes that he finds them "far from convincing," since molybdenum and tungsten from Hirsh's experimental setup are probably responsible for masking any lines from element 87. He concludes his note with the words:

> I feel his investigation does not help in the least to settle the question of the existence of a durable isotope of element 87.

Francium

Finally, in 1939, the real element 87, or eka-caesium, was truly discovered by an unknown French laboratory technician, Marguerite Perey. But Perey was no ordinary technician. She had the good fortune to become the personal assistant to Marie Curie and to be trained by her in the manipulation of radioisotopes. Eventually she would earn a degree, followed by a PhD and rise to the rank of professor of nuclear chemistry. It was in her PhD thesis where she reported the crucial experiments that show conclusively her discovery of the long sought for eka-caesium.

In retrospect, it is not difficult to see why most previous attempts did not amount to much, since all elements beyond atomic number 83, or bismuth, are radioactive, whereas previous attempts had mostly been conducted in nonradioactive minerals. At the time of Perey's work, several radioactive elements had been discovered in the previous fifteen years, mainly in France. They were polonium (84), radium (88), actinium (89), and protactinium (91). Any researcher hoping to discover element 87 had to be working in the field of radiochemistry if they stood any chance of being successful.

But even within radiochemical research there were a number of individuals who came close to discovering the element and again it is easier to understand why this was so in retrospect. In 1913, it was known that there were three main radioactive decay series that began with radium, thorium, and actinium. In addition, the radioactive displacement laws were already known, whereby the emission of α radiation resulted in the formation of an isotope with an atomic number of two units less than the decaying isotope, while β emission produced an isotope with one unit of atomic number higher. The former is explained by the fact that α particles have a charge of two. The latter is a little more involved, but is due to the fact that β emission is the result of a neutron in the nucleus of an atom decaying into a proton and a β particle. This amounts to the exchange of a neutron, of atomic number 0, for a proton of atomic number 1.

Consequently, from the early days of radiochemistry, it was apparent that element 87 would probably be formed from either the α decay of actinium (88) or the β decay of radon (86).

$${}_{89}\text{Ac} \rightarrow {}_{87}\text{eka-Cs} + {}_{2}\alpha$$

$${}_{886}\text{Rn} \rightarrow {}_{87}\text{eka-Cs} + {}_{-1}\beta$$

But oddly enough, the known isotopes of radon appeared to be just β emitters while those of actinium were only α emitters. Nevertheless, the possibility of an isotope decaying via both forms of deterioration was recognized. Eventually it turned out that eka-cesium was found as the result of the dual α and β decay of actinium and so we now turn briefly to the history of this particular element.

Actinium was discovered in 1900 by André Debierne, who was a colleague of the Curies in Paris. He called it actinium from the Greek ακτινοσ, or "ray," even though the observable radiation from this element is almost negligible. In 1908 Otto Hahn, then working at University College London, discovered what was first believed to be a new element and which was provisionally named mesothorium–2, or MsTh2. In fact, he had discovered another isotope of actinium, ^{228}Ac, in addition to Debierne's ^{227}Ac.

As a result, there were now two candidate isotopes of actinium that might be the source of eka-cesium. But no researcher pursued these possibilities or tried to detect whether they were actually α emitters. It was only in 1926 that Hahn decided to actively search for eka-Cs from his ^{228}Ac but failed to see any α decay. He realized that a similar search should be carried out with the other isotope, ^{227}Ac, but later claimed that he had lacked the skills required to purify actinium, which is precisely what Perey had and was able to utilize in her eventual discovery of the new element.

Another group of radiochemists who missed out on the discovery of eka-caesium were Meyer, Hess, and Paneth in Vienna. While conducting a survey of emissions from a number of isotopes they noticed that ^{227}Ac might indeed be an α emitter of very small intensity and reasoned that the radioactive decay series involving actinium might still be lacking some unknown elements. But this work was brought to an abrupt end by the outbreak of World War I.

As Kauffman and Adloff have claimed in their excellent article on the discovery of eka-cesium, the Vienna team had probably measured the direct α emission from ^{227}Ac but without being sure of it.[16] Had they known that this was the case they would of course have claimed the discovery of the new element. As mentioned earlier, the radioactive decay laws unequivocally imply that the emission of α radiation from actinium (89) produces an element with an atomic number of two units lower, namely element 87, or eka-Cs. But as late as the early 1930s, Marie Curie wrote in one of her books on radioactivity that actinium does not emit any observable radiation.

Perey

Marguerite Perey (fig. 7.3) joined Curie's laboratory, l'Institut du Radium, in 1929. From the beginning of her work Curie trained

FIGURE 7.3 Marguerite Perey. Image by permission of Emilio Segrè Collection at the Institute of Physics.

her in the skills of working with actinium, including concentrating the element when it was present in a mixture of rare earths that frequently included lanthanum. This task required repeated crystallizations and evaporations of compounds such as the oxalates of the metals. In addition, manipulations with actinium demanded working quickly because this radioactive series produces isotopes of thorium, radium, lead, bismuth, and thallium in quick succession. If actinium was to be studied for its own properties, all these daughter isotopes had to be quickly removed, especially as they produced their own α and β radiations.

Although Marie Curie died in 1934, Perey pursued the work on actinium, now under the joint direction of the discoverer of the element, Debierne, and also Irène Joliot-Curie, Marie Curie's daughter. Both of Perey's new mentors encouraged her to continue to purify and examine the radioactive properties of actinium. By now it was clear that ^{227}Ac had to be a β emitter since one of the daughter isotopes was ^{90}Th.[17] Still, nobody had observed this β emission from actinium itself.

In 1935 Hull, Libby, and Latimer in the United States thought they had detected such radiation. Perey's interest was aroused by this report, as she suspected that the radiation was probably due to some of the daughter isotopes. She was thus prompted to mount her own search for any β radiation coming directly from actinium. The work required that she concentrate a source of actinium and almost immediately (about one minute later) begin to look for its radiation, before it became swamped by that of its daughter isotopes.

In carrying out these procedures she discovered that during the first two hours the radioactivity increased rapidly, reached a plateau, and then increased again slowly. By extrapolating her graph to a time of zero she succeeded in obtaining a numerical estimate for the α and β radiation from pure actinium. It turned out that about 1 percent of the radioactivity was α decay and the remaining 99 percent was in the form of β decay. Perhaps the most crucial part of the observation was that the initial radioactive decay of actinium took place with a half-life of twenty-one minutes, a fact that she recorded on January 7, 1939, in her notebook, which survives to this day.

Piecing this information together, Perey deduced that the decay of ^{227}Ac occurs mainly via the formation of thorium, radium, and other daughter products, but that there is also a small, almost insignificant branching detour that could be taken to mean either the presence of a new element or possibly the existence of a yet unrecognized further isotope of actinium. Because Perey was confident that the α decay could not be attributed to daughter isotopes like thorium and radium, it had to be due to actinium itself. This meant that actinium might be decaying into element 87, given that α decay results in the lowering of atomic number by two units.

The rapid increase in β decay that she observed was not therefore due to actinium itself but perhaps to the decay element 87 (into ^{88}Ra

as it turned out). Of course, actinium itself shows β decay—in fact, 99 percent of its activity is due to this form of decay, as noted above. But since this process has a much longer half-life, it could be distinguished from the β decay from element 87. The chemical identity of the new element was confirmed by the fact that it could be precipitated with caesium salts. Caesium is an alkali metal in group 1 of the periodic table, as eka-caesium was expected to be since the time of Mendeleev's original prediction of the element. But of course none of this information was completely conclusive and there was still the question of whether a new isotope of actinium was giving rise to some of the observations on the decay of freshly prepared ^{227}Ac.

As a result, Perey showed due caution in announcing her discovery in a short article in the *Comptes Rendus de l'Académie des Sciences.* First, she made the case in favor of the possible new element:

> Dans l'hypothèse que ce radioélément est forme par émission de rayons α a partir de l'actinium, il occuperait la place 87 dans la classification périodique; pour le prouver nous avons cherche a constater l'analogie chimique de ce corps avec le caesium par syncrystallisation: le pérchlorat de caesium a été choisi pour cela, en raison de sa faible solubilité qui le distingue des perchlorates des métaux non alcalins, très solubles. En ajoutant du chlorure de caesium a l'eau mere et en précipitant par une solution de perchlorate de sodium, il se forme des cristaux qui entrainent l'activité: celle-si décroit éxponentiellment avec période de 21 minutes ± 1 (Perey, 1939, 89).[18]

A few lines later she wrote:

> Nous sommes donc amenée a pensér que cet element radioactif naturel, de période 21 minutes, a le numero 87 et derive, par rayonnement α, de l'actinium; soit que l'actinium possede un faible embranchement α, ou qu'il soit un mélange de deux isotopes se distintegrant l'un par rayonnement β, et l'autre par rayonnment α (Perey, 1939, 89).[19]

The paper was presented by Jean Perrin, who rather ironically believed that eka-caesium had been discovered earlier by one of his own collaborators, the Romanian physicist Horia Hulubei.[20] Perey initially named the new isotope actinium K since she could not be certain of its precise identity.

During the course of all this work, Perey had kept her two mentors, Debierne and Joliot-Curie, separately informed of her progress. Eventually, Joliot-Curie happened to mention to Debierne that Perey had discovered this new isotope or element as a result of her own suggestion. At this point Debierne is said to have fallen into a rage, having believed that he had been the sole director of Perey's work. One odd consequence of this dispute was that the two mentors could not agree as to which of them should share the credit with Perey, resulting in Perey being permitted to keep the discovery to herself.

When it became clear, as a result of further experiments, that Perey had in fact discovered a new element rather than a new isotope of actinium, she was invited to name her new discovery by Paneth, the head of the nomenclature commission of IUPAC. Perey suggested the name "catium" to mean cation, since the new element would form the largest cation in the periodic table. But this idea was opposed by Irène Joliot-Curie, who felt that speakers of the English language would mock the name because it sounded too much like the household pet, "cat." Perey then opted for the name "francium" to honor her country of France and suggested the symbol of Fa. Shortly afterward, the symbol was changed to Fr to be more in keeping with the manner in which other elements are symbolized, and that's how matters stand today.

Following Perey's discovery, she was encouraged by both of her mentors to obtain the degree she still lacked. Perey found herself in the remarkable situation of having enough material for a doctoral thesis and yet not possessing an undergraduate degree, which in turn meant that she could not even register for the doctoral process. Nevertheless, she quickly obtained various certificates in a number of subjects before finally defending her doctoral thesis in 1946. In 1949, she was appointed to a new chair in nuclear chemistry at the University of Strasbourg. Meanwhile, she had begun to suffer from radiation sickness as a result of her work with radioactive isotopes, and eventually died prematurely in 1975 at the age of sixty-five.

Perey had discovered the last naturally occurring element. All that remained to be found were elements 61 and 85, which would require their being artificially synthesized.[21]

Uses of Francium

The element francium is far too rare and far too radioactive to have any commercial applications. It does have some rather exotic scientific uses, however. Atoms of francium provide an excellent opportunity to examine many aspects of atomic and nuclear physics and even the theory that unifies the electromagnetic and the weak nuclear force into the electroweak force. These experiments have been dubbed "table-top physics" to distinguish them from high-energy physics carried out in huge particle accelerators spanning vast distances the size of cities.

Before describing this work in a general way it is necessary to look into the background physics a little. In the 1950s, Lee and Yang at Columbia University proposed that some elementary particles might violate parity. They suggested that such particles could distinguish between right and left, a seemingly odd notion. In 1957, one of their experimental colleagues at Columbia, Madame Wu, undertook a study of radioactive ^{60}Co and discovered to everybody's surprise that such parity violation was indeed a reality. When ^{60}Co atoms were placed in a magnetic field to polarize them and then allowed to undergo beta decay, they showed a greater tendency to emit electrons from their south poles than from their north poles. In the 1970s, the notion of parity violation was extended to stable, that is to say, nonradioactive atoms, and was observed in many atoms including caesium, and very recently, ytterbium.

Although the experiments with stable atoms give a much weaker effect than in the case of Wu's ^{60}Co, they do allow for the exploration of some new effects in the physics of the electroweak force. Moreover, the strength of the effect is roughly proportional to Z^3, the cube of the atomic number. Recent work on parity violation has tended to concentrate on atoms of caesium because of the simplicity of this system, with its one outer electron. This work has naturally suggested the use of francium, whose atoms likewise possess one outer shell electron and which are expected to have a parity violation effect eighteen times that of caesium.

Before parity violation experiments can be conducted on francium, the spectral transitions among the energy levels have to be experimentally determined. In addition, there has been much theoretical work aimed at calculating these energy levels and transitions. Such stringent tests of the theory have been proceeding for many years and agreement between theory and experiment has reached an accuracy of about 1 percent. Other experiments include the measurement of the "weak nuclear charge" of the nucleus by analogy with the more familiar electric charge in the electromagnetic force.

The next phase of experiments with francium are intended to move well beyond merely observing parity violation in atoms, since this feat has now been achieved in many different atoms. The real interest in francium is an attempt to measure more accurately than before a completely new predicted effect called the "anapole" moment, which is a fundamental aspect of the electroweak force arising through quantum mechanical effects.[22] It is called anapole, meaning not having to do with any particular kind of pole that appears in the electromagnetic force.[23]

Whereas an electromagnetic field can be expanded in terms of dipoles, quadrupoles, octopoles, and so on, the anapole associated with the electroweak force is simply not like that. It consists instead of a magnetic component that accompanies a so-called toroidal

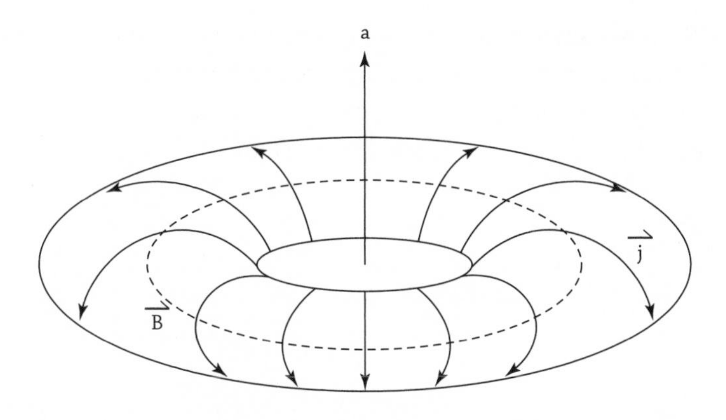

FIGURE 7.4 A toroidal current generates a nonzero anapole moment.

current, in which none of the current is actually found outside of the toroid shape (fig. 7.4).

Other aspects of the work have involved the development of new technologies to contain and focus beams of francium atoms as well as better methods to create francium isotopes of different kinds. The method used has been nuclear fusion reactions wherein beams of oxygen ions are accelerated to energies at which they can fuse with atoms of gold. Because of the lack of reactivity of gold, the most noble of the metals, the francium atoms that are created do not form any gold compounds and can be transported to a region where they can be "trapped."

The trapping work relies on the use of tunable lasers of various kinds as well as magnetic fields and resonant light. By the year 2002, such experiments had succeeded in maintaining as many as 300,000 atoms of francium in a magneto-optical trap, which then made possible the measurement of new spectroscopic transitions within the atoms.[24]

As Luis Orozco, one of the leading experts in this field, explains, plans are being laid to use different isotopes of francium in order to obtain mass variations while keeping all other features constant, which will lead to a deeper probing of the electroweak interactions.

The remarkable thing about all this work is that it consists of the exploration of the unification of two of the fundamental forces of nature, the electromagnetic and weak nuclear forces, but without the use of huge energies or particle accelerators.

Chapter 8

Element 85—Astatine

The story surrounding element 85 is one of the most complex and interesting among our seven elements (fig. 8.1). The various claims for its discovery reveal many of the nationalistic traits that we have seen in the case of other elements, most notably the controversy surrounding the discovery of hafnium, element 72.

But element 85 gives our study a greater depth than has yet been revealed by the already covered elements. What this story shows is that the nationalistic prejudices persist to this day in many respects and that the identity of the "discoverer" of the element very much depends on the nationality of the textbook that one might consult. It is also an element for which the majority of sources give an incorrect account in declaring Corson, MacKenzie, and Segrè as the true discoverers. The account I will detail owes much to the recent work of two young chemists, Brett Thornton and Shawn Burdette, whose 2010 article I have drawn heavily from.[1]

As in the case of many of the seven elements already surveyed, the view that Moseley's experimental demonstration of the concept of atomic number resolved all issues in a categorical fashion is once again shown to be highly misleading.

Early Claims for Element 85

The position of element 85 in the periodic table shows it to lie among the halogens. Not surprisingly, therefore, the early researchers believed that they would find the element in similar locations to other halogens such as bromine and iodine, namely in the oceans or

H																	He
Li	Be											B	C	N	O	F	Ne
Na	Mg											Al	Si	P	S	Cl	Ar
K	Ca	Sc	Ti	V	Cr	Mn	Fe	Co	Ni	Cu	Zn	Ga	Ge	As	Se	Br	Kr
Rb	Sr	Y	Zr	Nb	Mo	Tc	Ru	Rh	Pd	Ag	Cd	In	Sn	Sb	Te	I	Xe
Cs	Ba	Lu	Hf	Ta	W	Re	Os	Ir	Pt	Au	Hg	Tl	Pb	Bi	Po	**At**	Rn
Fr	Ra	Lr	Rf	Db	Sg	Bh	Hs	Mt	Ds	Rg	Cn		Fl		Lv		

La	Ce	Pr	Nd	Pm	Sm	Eu	Gd	Tb	Dy	Ho	Er	Tm	Yb
Ac	Th	Pa	U	Np	Pu	Am	Cm	Bk	Cf	Es	Fm	Md	No

FIGURE 8.1 Showing position of element 85, or as eventually named, astatine.

in sands washed up by oceans. Moreover, it was fully expected that the new element would behave like a typical halogen to form diatomic molecules and that it would have a low boiling point.

The first major claim for the discovery of the element was made by Fred Allison, the same researcher who also erroneously claimed that he had discovered element 87. And just as in the case of element 87, Allison claimed to have found the new element using his own magneto-optical method, involving a time delay in the Faraday effect, which is to say the rotation of plane polarized light carried out by the application of a magnetic field to any particular solution of a substance. Allison published articles in 1931 and 1932 claiming that he had observed element 85 and proposing to call it alabamium after Alabama, the state in which he worked.

In 1935, the American physicist H. G. MacPherson showed that Allison's findings were spurious and due to imperfections in his instruments rather than to the presence of a new element. Further refutations followed in quick succession.[2]

The next claim came from Rajendralal De, an Indian chemist working in Dacca, then part of British India and now in Bangladesh. De had trained in Germany with Hahn and Meitner in the 1920s and like Allison, used monazite sand for his research. After applying a number of chemical processes to the sand, De obtained a sublimate that he claimed to be element 85 and to which he gave the name of dakin after the city of Dacca, also spelled Dhaka. Later researchers dismissed De's claim on the basis of the powerful radioactivity of astatine, which would have prevented him or anybody else from safely handling the element in the manner he claimed to have done at the time.

Another person who had been involved in the search for element 87, Romanian Horia Hulubei, was also involved in the discovery of element 85. Indeed it appears that he may well have been the discoverer of naturally occurring astatine, as it was later called by the physicists who synthesized the element artificially. It is these physicists who are generally accorded with the discovery of the element.

Hulubei studied in France starting in 1916, returning to his native Romania after World War I had ended in 1918. In 1926, he came back to France to work with Jean Perrin and built an X-ray laboratory at the Sorbonne University. In 1928, they were joined by Yvette Cauchois, who built what later became known as the Cauchois spectrometer, which provided higher resolution spectra and made possible the study of weaker spectra than had previously been observed. Hulubei and Cauchois examined the radioactivity of radon in the hope of observing evidence of the presence of element 85. In a paper published in 1936, they claimed to have observed a line at 151 X-units, or siegbahns, precisely where the $K_{\alpha 1}$ line for eka-iodine was expected.[3] In 1939, they reported two further X-ray lines consistent with the presence of eka-iodine and the predictions from Moseley's law. These new experiments used higher resolutions than the earlier ones and included further checks and balances, which led to greater confidence in the authors' claims to having discovered the new element. In 1941, a former student of Hulubei and Cauchois, Manuel

Valadares, repeated the experiments with a stronger X-ray source after returning to his native Portugal. He then published his results, which also suggested the presence of eka-iodine.[4]

In 1942, additional scientists entered the discussion on the new element. Two women, Berta Karlik and Traude Bernert, working at the Institute for Radium Research in Vienna, reported the detection of α particles emanating from the radioactive decay of a radon isotope.[5] They also took this decay to indicate the presence of element 85 in part of a natural radioactive decay series. By this time the artificial synthesis of element 85, which is generally considered to be the definitive discovery of the element, had been conducted at Berkeley. The Austrian researchers were unaware of this fact, however, due to lack of communication during wartime.

In an article of 1944, Hulubei wrote a detailed summary of his work and that of others on element 85. This included a description of six X-ray lines that were thought to be due to natural radioactive decay producing the new element. He also appealed to the work of Karlik as providing support for his own findings. This time Hulubei went as far as to suggest a name for the new element, "dor," which he took from the Romanian word for "longing" in the sense of "longing for peace." This name represented an interesting shift away from naming elements in a nationalistic manner that had prevailed in the recent past.

As World War II drew to a close and some elements began to be produced artificially, it became important to decide on how elements should be named and who would have the right to give them new names. This task was taken up by the Austrian-born radiochemist Friedrich Paneth, who had fled from Berlin to the United Kingdom in 1936 after being dismissed from his professorship because of his Jewish origins. Paneth published an editorial in *Nature* magazine in 1947, which among other things would have the effect of depriving any discovery claims from Hulubei and Cauchois. As mentioned before, Paneth suggested that in cases in which an element had been given different names by competing groups, the naming rights should go to those who produced the element in a reproducible fashion. This meant that, in the case of element 43, the Noddacks' claim for masurium should be dismissed and should be replaced by technetium, as synthesized by Segrè and Perrier.

Paneth noted the claim by the Berkeley group for the synthesis of element 85 and also the fact that Karlik and Bernert had showed that it exists in natural sources. But he went on to state that what he called "former claims," without naming any particular researchers, had been disproved by the work of Karlik and Bernert. This is a rather crucial statement because it served to discredit the work of Hulubei and Cauchois, even though Karlik and Bernert had not actually addressed these claims whereas Paneth's statement implied that they had.[6]

Hulubei was understandably very concerned with Paneth's editorial and the implication that his work and that of Cauchois had been refuted. He responded by attributing Paneth's omission to the difficulties in communication during the war. He denied that Karlik and Bernert had refuted his research on element 85, adding the words, "contrary to what one would think after reading the expose of Mr. Paneth." Soon afterward Karlik finally did comment on Hulubei's work, claiming that the research had been insufficient to merit the discovery of element 85 because of the very small amount of element 85 in their sample, which would render likely some interferences from other elements in the X-ray spectra.[7]

Meanwhile, in response to Paneth's editorial, three Berkeley researchers claiming to have produced element 85 artificially—Corson, MacKenzie, and the previously mentioned Emilio Segrè—proposed the name "astatine" from the Greek *astatos,* or unstable. The authors had not been aware of the claims from Hulubei and Karlik but had delayed proposing a name for the element because of the continuing claims for alabamine by Allison and his supporters. Furthermore, Paneth, who was by now the chair of the committee of the International Union of Chemistry, approved the name of astatine in 1949, thus further lending his support to the American claim.

According to the analysis of Thornton and Burdette, there is no doubt that three teams of researchers can claim to have discovered element 85.[8] First of all, they state that:

> Unlike other flawed studies with X-ray spectroscopy, Hulubei and Cauchois indisputably had element 85 in their samples. The only uncertainty is whether their instrument was sensitive enough to distinguish the spectral lines of element 85.[9]

One additional argument they offer for this claim is that, in the 1930s, Hulubei and Cauchois were able to clearly detect the L_α line for the element polonium, which has a 500-fold lower transition intensity than the lines they claimed to have seen in the case of element 85. Moreover, they add that the experiments carried out by their Portuguese student, Valadares, would have tripled the intensity in the claimed X-ray lines for element 85 because he used a radon source, which is three times more powerfully radioactive.

The reasons why Hulubei and Cauchois have never received much credit for their work have already been mentioned. They include Paneth's disparaging words to the effect that "other work" on element 85 had been refuted even though Hulubei and Cauchois's work had not. In addition, Thornton and Burdette attribute the lack of credit to the fact that Hulubei in particular had falsely claimed the discovery of element 87 and that he had definitely been wrong in that case. They propose that this earlier error caused others to doubt Hulubei, even though he had detected element 85.[10]

Helvetium and Anglohelvetium

In 1940, the Swiss physicist Walter Minder (1905–1992), claimed to have observed an extremely weak β decay of radium A. For this purpose he connected a couple of ionization chambers with an electrometer. He also believed that his chemical tests confirmed the analogies of this element with iodine. Minder named it helvetium and gave it the symbol Hv, after the Latin name for Switzerland. *Nature* Magazine reported Minder's findings in an abstract by announcing that he had succeeded in isolating element 85 and that he had done so from the decomposition of the radioactive element actinium. The abstract also noted that Minder had named his new element helvetium to honor his own country. It continued by expressing the hope that further details would soon be available, adding that the *London Evening News* had remarked,[11]

> It is odd to learn today, in the midst of war, that a patient Swiss chemist has succeeded at last in isolating the elusive chemical element '85.' It is still odder that in the long view of history a discovery of that sort may rank above all the perils and victories of these days.

Then in 1942, Minder with his British colleague, Alice Leigh-Smith,[12] surprisingly repeated the announcement of the discovery of eka-iodine, this time calling it anglohelvetium, a combination of Anglia (Latin for England) and Helvetia. But again others could not replicate these claims and not much was heard from these researchers again, at least in the context of the discovery of missing elements.

The Usually Acknowledged Discovery of Element 85

The discovery of element 85 was made by three Berkeley scientists, Dale Corson, Alexander MacKenzie, and Emilio Segrè, in 1940 (fig. 8.2). Using a 60-inch cyclotron built by Ernest Lawrence, the three scientists bombarded a target of bismuth, element 83.[13] This element is rather useful in this context because of its having just one single isotope, with mass number 209, a feature that greatly simplifies the

FIGURE 8.2 One of the codiscoverers of astatine, Dale Corson. By permission from Emilio Segrè Collection at the Institute of Physics.

analysis of products. The bombarded bismuth target displayed a number of forms of radiation, including the emission of α, γ, X-rays and even low energy electrons, all exhibiting the same lifetime of about seven-and-a-half hours. Through a series of analyses the authors were able to identify the substance causing some of these radiations, with element 85 changing into polonium via K-electron capture.

$$^{211}_{85}\text{At} + {}_{-1}\text{e} \rightarrow {}^{211}_{84}\text{Po} + 90 \text{ kilovolt X-ray}$$

Interestingly, in the article announcing their discovery, they also remarked about the possible existence of naturally occurring element 85 and cited the earlier work of Minder in Switzerland, as well as Hulubei and Cauchois in Paris, both of whom had claimed to have observed the element.

They also mentioned the work carried out with Hamilton and Soley in which element 85 was concentrated into the thyroid glands of some guinea pigs, showing similar excretion to that of iodine, which occurs above element 85 in the periodic table. Nevertheless, the chemical experiments of Corson et al. revealed that the properties of element 85 are more similar to those of neighboring element 84 or polonium than they are to iodine. For example, element 85 precipitates as a sulfide and is precipitated by zinc in sulfuric acid, both of which are reactions that are characteristic of a metal rather than a nonmetal such as iodine.

General Aspects of Astatine

Element 85 has the dubious distinction of being one of very few solid elements that has never been obtained in any amount large enough to be visible to the naked eye. It is also estimated that if a visible sample were ever produced, it would immediately vaporize away due to the heat generated by the emitted radioactivity. As a result of these properties the bulk behavior of astatine, such as its melting and boiling points, its color, and the degree to which it may be a metal can only be estimated theoretically.

Based on melting point trends among the halogen elements, the value for astatine is predicted to be 302°C, although there is some controversy surrounding this work, as there is around the predicted melting point of 337°C.[14] Another controversy concerns the apparently simple question of whether diatomic molecules of At_2 occur as they do in the case of all the other halogens.[15] The color is expected to be very dark and most probably black on the basis of the trend among the halogen group, to which astatine belongs. This is because fluorine is almost colorless to yellow, chlorine is green, bromine brown, and iodine a violet color.

In 1943, three years after astatine was first synthesized artificially in a nuclear reactor, it was discovered that the element occurs naturally in miniscule amounts in the earth's crust. In fact it is the single rarest naturally occurring element, with a total of just 1 oz or 28 grams at any given time. About thirty isotopes of the element have been synthesized or found to occur naturally, the longest lived of which is ^{210}At with a half-life of 8.1 hours.

Taken all together these facts about the element contribute to its almost complete lack of applications. One exception has been an ongoing exploration of the potential uses of ^{211}At in radiotherapy. The isotope is an α particle emitter with a convenient half-life of 7.2 hours. Like the element above it in the periodic table, iodine, astatine has a tendency to be metabolized in the thyroid gland and could therefore be used to monitor medical conditions involving the thyroid and the throat area in general. In addition, the short-range nature of the α emission of this isotope suggests that it could be used to treat cancers in all parts of the body while reducing the risk to neighboring tissue that is often a problem in the use of other more established radio-therapeutic isotopes. And if that were not promise enough, ^{211}At does not produce any harmful β radiation as do many other isotopes currently used in radio-medicine.

But although these therapeutic potentially attractive properties have been explored for more than thirty-five years, problems concerning the safe delivery of ^{211}At to human subjects, as well as issues relating to the ready production of the isotope, continue to delay the in vivo implementation of this rarest of all elements.[16]

Chapter 9

Element 61—Promethium

H																	He
Li	Be											B	C	N	O	F	Ne
Na	Mg											Al	Si	P	S	Cl	Ar
K	Ca	Sc	Ti	V	Cr	Mn	Fe	Co	Ni	Cu	Zn	Ga	Ge	As	Se	Br	Kr
Rb	Sr	Y	Zr	Nb	Mo	Tc	Ru	Rh	Pd	Ag	Cd	In	Sn	Sb	Te	I	Xe
Cs	Ba	Lu	Hf	Ta	W	Re	Os	Ir	Pt	Au	Hg	Tl	Pb	Bi	Po	At	Rn
Fr	Ra	Lr	Rf	Db	Sg	Bh	Hs	Mt	Ds	Rg	Cn		Fl		Lv		

La	Ce	Pr	Nd	**Pm**	Sm	Eu	Gd	Tb	Dy	Ho	Er	Tm	Yb
Ac	Th	Pa	U	Np	Pu	Am	Cm	Bk	Cf	Es	Fm	Md	No

FIGURE 9.1 Updated periodic table showing position of element 61, eventually named promethium.

The last of our seven elements to be isolated was element 61, which is also the only rare earth among the seven (fig. 9.1). The problem with rare earths, which are 15 or even 17 in number depending on precisely how they are counted, is that they are extremely similar to each other and as a result are very difficult to separate. When the periodic table was first discovered in the 1860s only two or three rare earths even existed (fig. 9.2). As more of them turned up it became increasingly difficult to place them in the periodic system.

Early Claims

Just like with all the other seven elements in our story, there were many false claims to its discovery. Moreover, the early claims must have seemed very plausible at the time because they appeared to draw support from X-ray evidence and Moseley's law. Just like the priority dispute involving hafnium that took place in the early 1920s, the case of element 61 also involved an international controversy. This time one cannot entirely blame the aftermath of the Great War, as the two opponents consisted of Italians and Americans, with much of the scientific chicanery taking place, as was usual for the time, in the pages of London's *Nature* magazine.

But even though both sides of the priority dispute appealed to X-ray data and Moseley's law, it turned out that neither side was right. In their own way, each side was working in complete delusion, since element 61 is highly radioactive and unstable, does not occur naturally on Earth, and could only be isolated in minute quantities by artificial means when such methods became sufficiently developed in the 1940s.

Let us start at the beginning. In 1902, the Bohemian[1] rare earth chemist Bohuslav Brauner was the first to suggest that an element lying precisely between neodymium and samarium remained to be discovered. He gave talks in his native Bohemia and published articles in some fairly obscure journals, all of which meant that few chemists in the wider arena became aware of his work. In 1927, during the height of the priority debate over element 61, Brauner felt compelled to assert his priority, not over the discovery of the element but regarding his 1902 prediction that such an element should even exist between neodymium and samarium (fig. 9.2). This letter to *Nature* is also interesting because it again highlights the fact that Moseley's method is not quite as powerful as often portrayed.

But first a little background on Brauner. Beginning in the 1870s, Brauner worked on chemical substances that supported the validity of Mendeleev's periodic law. In 1881, he began a lively

Das periodische System der Elemente.

Gruppen	I.	II.	III.	IV.	V.	VI.	VII.	VIII.
Reihen	(RX_7) R_2O	(RX_6) R_2O_2	(RX_5) R_2O_3	RH_4 R_2O_4	RH_3 R_2O_5	RH_2 R_2O_6	RH R_2O_7	(R_2H) (R_2O_8) } Verbindungsformen
1. 2.	1 Li Li 7	 Be 9	 B 11	 C 12	 N 14	 O 16	 F 19	
3. 4.	23 Na K 39	24 Mg Ca 40	27 Al Sc 44	28 Si Ti 48	31 P V 51	32 S Cr 52	35.5 Cl Mn 55	 Fe 56, Co 59, Ni 59, Cu 63
5. 6.	(63 Cu) Rb 85	65 Zn Sr 87	69 Ga Y 89	72 ? Zr 90	75 As Nb 94	78 Se Mo 96	80 Br ? 100	 Ru 104, Rh 104, Pd 106, Ag 108
7. 8.	(108 Ag) Cs 133	112 Cd Ba 137	114 In La 139	118 Sn Ce 141.6	120 Sb Di 146.7	126 Te Tb 148.8 ?	127 J Sm 150 ?	 ? 152, ? 153, ? 154, ? 156
9. 10.	156 ? ? 170	158 ? ? 172	? 159 Ya? Yb 173	162 ? ? 177	166 Er? Ta 182	167 ? W 184	? 169 Tm? ? 190	 Os 193[1]), Jr 193, Pt 195, Au 197
11. 12.	(197 An) ? 221	200 Hg ? 225	204 TI ? 230	207 Pb Th 234	210 Bi ? 237	? 214 Ng ? U 240	219 ? ? 244	

1) Aus der Dampfdichte des Os O4 (Deville und Debray, Ann. chim. phys. (3) 56, 476) ergiebt sich die Zahl 193 als Atomgewicht des Osmiums.

FIGURE 9.2 Brauner's periodic table of 1882 with a homologous accommodation of the rare earth elements. *Chem. News*, 58, 307: At this stage there is no hint of an element between Nd and Sm. In fact, Nd had not even been discovered.

correspondence with Mendeleev, and a strong personal friendship developed between them. Among other things, Brauner demonstrated that beryllium in its compounds is bivalent and not trivalent, and thus confirmed the accuracy of Mendeleev's correction of the atomic weight of beryllium according to the periodic law. Brauner's studies on the rare earth elements and the determinations of their atomic weights were of particular importance, causing Mendeleev to remark that Brauner was one of the first chemists to confirm the conclusions from the periodic law with regard to cerium. In 1900, Brauner proposed that the rare earth elements be placed in a distinctive "interperiodic" group immediately after lanthanum. His fundamental idea was eventually corroborated by discoveries in atomic structure. At Mendeleev's request, Brauner also wrote the large section "Rare-Earth Elements" for the seventh edition of Mendeleev's famous book *Osnovy khimii* or *The Principles of Chemistry.*

At the time of his 1927 letter to *Nature,* Brauner was seventy-two years old. He began by congratulating the American researchers for their discovery of illinium, their name for element 61. He continued by drawing attention to the start of the American paper in which it was claimed that:

> there was no theoretical grounds for supporting that eka-neodymium [*sic*!] might exist until Moseley's rule showed that element number 61 was still to be identified.

Brauner then reminded readers that he had devoted almost all his scientific life, since 1877, to the study of the rare earth elements and their role in the periodic table of the elements. Brauner comments on how he had arrived at the conviction that the gap between the atomic weights of neodymium and samarium was abnormally large and amounted to 6.1 units of atomic weight. This value, he wrote, was larger than between any two elements in the periodic table and of the same order of magnitude as the gap between molybdenum and ruthenium (5.7), between which occurs the element eka-manganese.[2] Moreover, he points out that it is also of the same order of magnitude as the gap between osmium and tungsten (6.9), between which falls the element dvi-manganese.[3]

Brauner's point is clear. Given that these two elements below manganese were fully anticipated, even by Mendeleev, it is clear that the gap in atomic weight between neodymium and samarium is pointing to another new element. One does not need Moseley's X-ray method or Moseley's law in order to make such a prediction. In the same paragraph Brauner makes a remark that is perhaps more interesting for all the elements discussed in the present book. He states that he had predicted the discovery of seven elements, with atomic numbers 43, 61, 72, 75, 85, 87, and 89. This is quite remarkable given that even Moseley, armed with his experimental method and his law, was not able to reach this conclusion to anything like the degree of accuracy achieved by Brauner, on purely chemical grounds some twelve years prior. Moseley himself could only predict the existence of elements 43, 61, and 75 with any confidence.

The only difference between Brauner's sequence of seven elements and the seven elements that are the subject of the present book is his inclusion of element 89 and his omission of element 91. In fact the isolation of element 89, which was eventually called actinium, had not been definitively settled when Brauner made his predictions in 1902.[4] The matter was not resolved until 1904 and only partly so. Brauner's only "mistake," therefore, is failing to predict element 91, the discovery of which was the subject of chapter 3.

Even here it is somewhat understandable why Brauner failed to predict this element. Unlike in other cases where he could search for large gaps between the atomic weights of consecutive elements, the atomic weight of element 91 presents a rare case of a pair reversal of which there are only five in the entire periodic table.[5] The atomic weights (modern values) of the three consecutive elements thorium (90), protactinium (91), and uranium (92) are 232.03, 231.035, and 238.02. On the other hand, perhaps Brauner could have predicted an element between thorium and uranium regardless of the pair reversal since the gap between thorium and uranium consists of almost exactly 6 units on the atomic weight scale.

Finally, Brauner presents another argument, which he had mentioned in lectures to the Bohemian Chemical Society and to the St. Petersburg Academy, and which did not depend on gaps between atomic weights of elements in the periodic table. Brauner wrote:

> On arranging the true hydrides (in which hydrogen is negative towards the positive metal) of the elements of the 8th series of the periodic system according to the order of their atomic weights, we find the following remarkable regularity of the composition of those peculiar compounds:
>
> CsH_1, BaH_2, LaH_3, CeH_4, PrH_3, NdH_2, XH_1, SmH_0.
>
> As samarium does not combine with hydrogen there must exist between neodymium and samarium an unknown element which forms the hydride XH_1—and this element is illinium. My speculation has not proved futile.

An Italian Claim

In 1924, a team from Florence, Italy, believed that they had found element 61. Luigi Rolla and Lorenzo Fernandes did not publish their findings until 1926, however. Writing in an Italian chemical journal in 1926, Rolla and Fernandes described the experiments on Brazilian monazite sand that they had begun in 1922. They wrote that their examination of the L series of the resulting X-ray spectra yielded no positive results. However, they also claimed that experiments on the K series of the same samples, carried out by Professora Signorina Brunetti, yielded the characteristic frequencies for element 61. They explained that on first obtaining these results in June of 1924, the quantity of the material they possessed was so small that they preferred to deposit the X-ray photographs in a sealed parcel at the Accademia dei Lincei rather than publish their results:

> Il materiale che noi avevamo a disposizione era però in cosi piccola quantità, che non ritenemmo scrupoloso pubblicare allora nostre ricerche, e inviammo al R. Accademia dei Lincei (nel giugno 1924), un plico suggellato contenente i nostri risultati e le fotografie dello spettro del quale si è parlato.[6]

They went on to describe how they had recently resumed their work on these samples and had carried out what seems today to be a staggering three thousand crystallizations in order to purify their suspected new element. The outcome of this work seems to have been a new series of spectra that showed the same spectral anomalies that had led them to suspect the presence of an element in the first place. The authors then seemed to concede that they were going into print because of the recent announcement, from the United States, by Harris, Hopkins, and Yntema, who in the meantime had published their claim to having discovered the element they were calling illinium after their home state of Illinois.

In recent years, a group of Italians have written an interesting and detailed account of the work of Rolla and Fernandes.[7] Although they do not attempt to rehabilitate the work of their compatriots, they seem unable to resist giving at least some credit for the discovery of element 61 to both the Florentine and American teams:

> Noi ipotzizziamo che I campioni analizzati in Firenze ed in America potessero contenere tracce minime di questo elemento.[8]

Charles James and B. Smith Hopkins

The next two authors who claimed to have discovered element 61 will be examined in parallel even though they worked independently, because their stories are somewhat entwined. Although the majority of sources discuss the contribution of Smith Hopkins, very few of them mention his contemporary Charles James. This bias is partly due to the fact that Smith Hopkins published his claim in a more prominent journal, the *Journal of the American Chemical Society*, and perhaps also because he proposed a name for his element, something that the more retiring James did not do.[9]

Here is the way in which Clarence Murphy, who has made a special study of the work of James and the discovery of element 61, begins his article:

> The history of the search for and discovery of Element 61 is one of the most complex and confused of any of the elements in the periodic table. Certainly no element has been "discovered" and named more times than 61. At least seven claims for discovery were made and 61 has been named at various times illinium, florentium, cyclonium, and promethium. The story of element 61 is also intimately connected with the development of the understanding of atomic structure and of the Periodic Table, and of advances in science and technology in the late 19th and early 20th centuries. The story involves Roentgen's discovery of X-rays and Moseley's use of X-ray spectra to determine atomic numbers. It involves the more than one hundred-year effort to separate the rare earths and to find a place for them in the Periodic Table. Finally it involves the development of ion-exchange chromatography and research on the atomic bomb during World War II. Element 61 was named prometheum in 1946 by its discoverers Coryell, Marinsky, and Glendenin after the Titan Prometheus, who stole fire from the gods and was sentenced to eternal torment for the crime, as a warning that atomic energy could be the savior or the destroyer of humankind. The spelling was later changed to promethium by IUPAC. (quoted with permission from *Bulletin for the History of Chemistry*).

I will start with Charles James, since he seems to have initiated his work on the rare earths, and element 61 in particular, before Smith Hopkins did. James was born in England and showed an interest in chemistry from a young age. While still in high school he wrote several letters to Ramsay, the University College chemist and discoverer of all but one of the noble gases. James began to search for element 61 following a letter from Sir William Ramsay in 1912, in which the latter pointed out that there were large gaps between the atomic weights of certain apparently consecutive elements in the periodic table. In particular, Ramsay singled out a suspected gap between the elements neodymium and samarium.[10]

At the even earlier date of 1908, James was corresponding with another British knight, Sir William Crookes, who was apparently analyzing some rare earth samples for James. James's connection with the British chemical intelligensia is further highlighted by letters to and from Henry Moseley. In one of these letters, dated 1912, before Moseley's major breakthrough concerning atomic number, Moseley is asking James to send him a sample of thulium. Clearly, James's search for element 61 did not need to wait for Moseley's definitive statement that a gap existed for this element.

The paper in which James made his claim for the discovery of element 61 appeared in the *Proceedings of the National Academy of Science* and not in a mainstream chemistry journal. It is coauthored with his then graduate student at the University of New Hampshire, Herman Fogg, as well as James Cork of the Department of Physics at the University of Michigan. It was Cork who carried out the vital X-ray spectral measurements on the samples provided by James and Fogg. Here is a key passage in the paper:

> In making measurements of the wave-lengths of the X-ray K emission lines for the rare-earth elements, very faint traces of lines corresponding to the K series of the element of atomic number 61 appeared on the plate with certain specimens of samarium (62) and neodymium (60).

James's research program to search for element 61 had begun in 1923 and he had used a variety of minerals including gadolinite, ytterspar, and monazite. The work involved innumerable fractionations using many techniques that James had developed himself over many years and that were adopted by many other rare earth researchers. After uncovering what he termed "traces of element

61" James attempted to confirm the presence of the element using a large quantity of monazite, the ore that seemed to yield the most promising results. Finally, the fraction thought to contain the new element was sent to James Cork, an expert in making X-ray measurements on trace elements.

As it turned out, this final step involving cooperation with another university seems to have caused delays, as Cork took some time to get back to James with the results. In the meantime, James's competitor Smith Hopkins had already published his claim to the discovery of the element on the basis of what turned out to be less substantial X-ray evidence. Returning to James's and Cork's claim, here are the supporting data that they published on element 61. Using some lines in what they took to be the X-ray L spectrum for the element, and after confirming their belief that these lines fell between those for elements 60 and 62, they published the following results:

2.289, 2.279, 2.078, 2.038, 1.952, 1.799 and 1.725 X.U.

To put this into perspective, the authors also said that there were about twenty lines in the L series for each element, whereas only seven of them were fairly strong.

The quite remarkable thing is that in 1949, two years after element 61, or promethium as it became, was synthesized, a team from the Oak Ridge National Laboratory set out to determine the L spectrum of the "real element" and published their results in the *Physical Review*.[11] Moreover, they compared the spectrum of the element synthesized at Oak Ridge with the spectrum reported twenty-three years earlier by James, Fogg, and Cork. The result of this comparison as well as a comparison with the lines obtained independently by Smith Hopkins were presented in a table below (fig. 9.3).

In spite of what seems to be a rather close coincidence between their own results and those of James, Fogg, and Cork, there is no hint of a comment one way or the other by the Oak Ridge team, as to whether or not the measurements support the earlier claims for the discovery of element 61.

Line	Cork, James, Fogg. 1926	Harris, Yntema, Hopkins. 1926	'real' element 61 1949
$L_{\alpha 1}$	2289		2287.9 ± 0.4 XU
$L_{\alpha 2}$	2279	2278.1 ± 3.0	2277.5 ± 0.3
$L_{\beta 1}$	2078	2077	2075.4 ± 0.4
$L_{\beta 3}$	2038		2037.9 ± 0.4
$L_{\beta 2}$	1952		1951.8 ± 0.6
$L_{\gamma 1}$	1799		1795.2 ± 0.9

FIGURE 9.3 Based on W. F. Peed, K. J. Spitzer, and L. E. Burkhart, The L Spectrum of Element 61, *Physical Review* 76, 143–144, 1949.

$L_{\alpha 2}$	2288	2289
$L_{\alpha 1}$	2278	2279
$L_{\beta 1}$	2076	2078
$L_{\beta 3}$	2038	2038
$L_{\beta 2}$	1952	1952
$L_{\gamma 1}$	1796	1799

FIGURE 9.4 Comparison of X-ray lines reported by Cork, James, and Fogg, with those for "real" element 61.

Let us assume that we consider the upper estimates of the 1949 measurements by including the estimated errors as published by the authors in the table and that we round our values to coincide with the accuracy reported by Cork, James, and Fogg. Here then is how the two sets of values would match up together (fig. 9.4).

Notice that two of these comparisons now represent an exact match while a further two show a difference of just one unit in about 2300. [12] This may be why Clarence Murphy writing as recently as

2006 also seems to support the claim that James might have actually discovered element 61.[13]

As mentioned above, Cork, James, and Fogg's paper was delayed because of the time that Cork took in sending his colleagues his measurements. But another perhaps even more significant factor conspired to weaken their claim. When James was almost ready to publish his own findings, he received a request from the editor of the *Journal of the American Chemical Society* to referee an article on the discovery of element 61 by Smith Hopkins at the University of Illinois. Rather than trying to find fault in this paper, James recommended publication to the editor but thereby prevented the possibility of his own submission, which would in all probability also have been made to this same journal.[14] In order to avoid any conflict of interest James therefore submitted his own paper to *Proceedings of the National Academy of Science.* [15] In any case, according to current estimates of the abundance of element 61 in the Earth's crust, it appears that James cannot have isolated the element, regardless of Murphy's attempt to rehabilitate James's work.[16]

The 1947 Discovery of the "Real Element 61"

As mentioned earlier, the discoverers of this element did not deliberately set out to synthesize it. Rather, they were engaged in trying to identify various products of nuclear reactions that were being explored in the Manhattan Project and the scientific research this project gave rise to.

The method they used to do so was called ion exchange chromatography. As in all forms of chromatography, the technique involves separating the components in a mixture by using a solvent and some medium such as a piece of blotting paper or a column of some special material that leads to different rates of movement of the components in the mixture.

Although ion exchange chromatography had begun as early as the 1850s, it was vastly improved during the Manhattan Project because of the need to separate numerous rare earth isotopes, including those of uranium and plutonium. The new feature was the use of special absorbents that could latch onto charged ions of the rare earth isotopes, which would then show different rates of elution along the selected column.

When Marinsky and Glendenin, a postdoctoral and graduate student, respectively, commenced their work at MIT, they took up a finding made in the Manhattan Project that suggested two unidentified fission products were present among some rare earth elements following the irradiation of a neodymium target. What was clear from these earlier studies was that the unidentified isotopes consisted of either praseodymium, neodymium, or element 61.

The ion exchange chromatography analysis by the pair from MIT made use of a synthetic organic cation exchanger called Amberlite IR, which consisted of a sulphonated phenol-formaldehyde compound. Experiments indicated that the order of elution was inversely related to the atomic number of the isotope in question. This can be seen in figure 9.5, reproduced from their article, in which elements 59, 58,

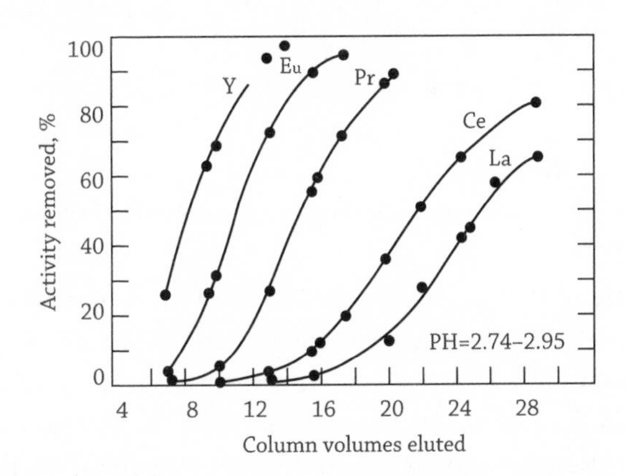

FIGURE 9.5 Ion-exchange chromatography and elution curves for various elements. Reprinted with permission of the American Chemical Society. From J. A. Marinsky, L. E. Glendenin, C. D. Coryell, The Chemical Identification of Radioisotopes of Neodymium and of Element, *Journal of the American Chemical Society, 69*, 2781–2785, 1947.

and 57, or praseodymium, cerium, and lanthanum, respectively, are eluted in such a way that the higher atomic number isotope of 59, or praseodymium, is the first to be eluted, followed by cerium (58), and then lanthanum (57).

In a separate experiment containing only elements 59, 60, and 61 (the praseodymium group) in addition to yttrium, the various isotopes in question produced the elution peaks shown in fig. 9.6. The peak at a volume of 1.6 liters was assigned to the element yttrium. Of the remaining four peaks those corresponding to 2.8 liters were assigned to neodymium (element 60), while the very tall peak at 3.6 liters was assigned to element 59, or praseodymium. Arguing by analogy with the previous diagram, and the inverse relationship between order of elution and atomic number, it became clear that the second tallest peak, corresponding to an activity of 17 units and a volume of about 2.6 liters, was due to an isotope of the new element number 61 (fig. 9.6).

In fact, two isotopes of element were discovered in these experiments at MIT. The first was an isotope with a half-life of 3.7 years corresponding to the peak just described, and assigned to a mass number of 147, while further work revealed a shorter-lived isotope of mass number 149 and with a half-life of forty-seven hours.

A Little Detour on Configurations of Atoms

The nature of the rare earths can be more easily understood from the perspective of the electronic configurations of their atoms of elements and from a knowledge of how their electron shells are occupied as one moves across the periodic table. Trying to understand this issue also involves taking a brief detour to examine the difference between the original 8-column short-form periodic tables (fig. 9.2), and the now almost ubiquitous medium long-form or 18-column table (fig. 9.1) and the long-form or 32-column periodic table (fig. 9.7). The move from the short form to the medium/long-form table involves removing the transition, or d-block, metals from the

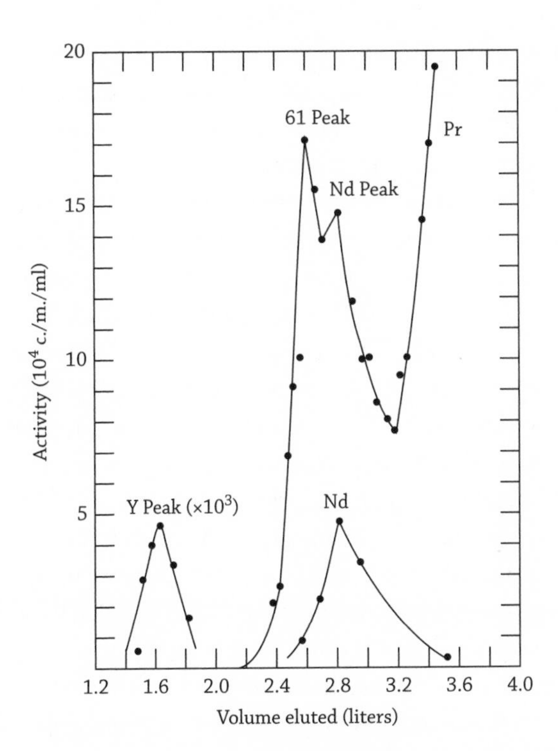

FIGURE 9.6 Elution curve of intermediate rare earth fraction. Reprinted with permission of the American Chemical Society. From J. A. Marinsky, L. E. Glendenin, C. D. Coryell, The Chemical Identification of Radioisotopes of Neodymium and of Element, *Journal of the American Chemical Society, 69,* 2781–2785, 1947.

main body of the table and placing them into a central block. This also reflects better the fact that the similarity among transition elements such as titanium, zirconium, and hafnium, for example, is greater than among the other elements that occur in the fourth column of the short-form periodic table, namely carbon, silicon, germanium, tin, and lead.

Similarly, one could in principle place the first few rare earth elements such as lanthanum, cerium, and praseodymium among the transition elements. It was soon realized, however, that it would be advantageous to separate them out into a distinct series. This is done by removing the entire rare earth block (or f-block). In the medium-form table, the rare earth block is placed at the foot of the main body of the table as a sort of footnote or afterthought. The way to avoid this is to move to the long-form table (fig. 9.7), which has thirty-two columns and in which the rare earths take up their natural positions so that there are no interruptions in the sequence of increasing atomic numbers that one finds in the medium-long form table.

But what bearing do electronic configurations have on this issue? The answer lies in the fact that as one crosses the transition metal block of elements, the new electron that differentiates each atom from the previous one, with a few exceptions, is added to the penultimate rather than the ultimate or outermost shell. For example, the configurations of scandium, titanium, and vanadium are: [Ar] $3d^1$ $4s^2$, [Ar] $3d^2$ $4s^2$ and [Ar] $3d^3$ $4s^2$. Meanwhile, in the case of traversing the rare earth block, the electron that differentiates the atom of each element as one moves across the table is occupying a shell that is even further from the outer-shell, namely two shells from the outer-shell.

The chemical result of these subtle electronic effects is rather profound. On moving across a short period that involves only the so-called main-group elements, such as beryllium, boron, carbon, and nitrogen, one sees a large difference in chemical as well as physical properties. Here the differentiating electron is entering an outermost shell and, given that the properties of atoms are governed by the number of electrons in the outer-shell, we observe a large variation of properties.

As we move across the first transition series, from scandium through titanium and on to vanadium, for example, the differentiating electron occurs in the penultimate shell, with the result that the variation in properties is less pronounced than across a series of main-group elements. Finally, when it comes to crossing the rare earth series, the variation in properties is even less pronounced, to the point that the elements are almost identical as a result of the entry of successive differentiating electrons at a distance two shell in from the outer shell. Recall again that properties are governed by

H																															He
U	Be																									B	C	N	O	F	Ne
Na	Mg																									Al	Si	P	S	Cl	Ar
K	Ca															Sc	Ti	V	Cr	Mn	Fe	Ce	Ni	Cu	Zn	Ga	Ge	As	Se	Br	Kr
Rb	Sr															Y	Zr	Nb	Mo	Te	Rs	Rh	Pd	Ag	Cd	In	Sn	Sb	Te	I	Xe
Ca	Ba	La	Ce	Pr	Nd	Pm	Sm	Eu	Gd	Th	Dy	Ho	Er	Tm	Yb	Lu	Ni	Ta	W	Re	Os	Ir	Pt	As	Hg	T1	Pb	Bi	Ps	At	Rs
Fr	Ra	As	Th	Ps	U	Np	Pu	Am	Cm	Bk	Cr	Es	Fm	Md	No	Lr	Rf	Db	Sg	Bh	Sh	Mt	Ds	Rg	Cn	113	F1	115	Lv	117	118

FIGURE 9.7 Long-form periodic table.

the number of outer-shell electrons. For the rare earths, not just one but two outermost shells are the same.

Mendeleev was operating in complete ignorance of all this information on electron configurations and how they influence chemical behavior. In fact, at the time of Mendeleev's finding, the electron had not yet been discovered. Mendeleev more or less gave up on the problem of how to place the rare earths into the periodic system and asked the Czech chemist Bohuslav Brauner to take over this onerous task.

Separation of Rare Earths

The classic method for separating the rare earths involves carrying out repeated fractional crystallizations, sometimes numbering as many as several hundred or even thousands of such tedious operations. Nevertheless, the chemists of the nineteenth century as well as the first half of the twentieth century persisted in these tasks and managed to isolate all but one of the rare earths, namely element 61. The discovery of this final element had to await the development of a new technique of separation, called ion-exchange chromatography, as we have seen.[17]

In addition, the classical methods of separation failed because there is simply not enough of element 61, if any, in the Earth's crust. And even if there ever was any of the element present, it has long ago decayed into other elements, as a result of its instability and the short lifetimes of all its isotopes.

The true discovery of element 61, therefore, also required the artificial synthesis of elements as was the case for technetium and astatine in previous chapters. Thus, it was the conjunction of the production of synthetic elements coupled with the improved techniques of ion exchange chromatography that finally allowed the periodic table of the first 92 elements to be completed when the gap at element number 61 was plugged in 1947.[18] It is also interesting to realize that the researchers concerned with this work were not deliberately setting out to form element 61. It was more a case of experiments producing a number of new isotopes and attempts to characterize these products that led the chemists and physicists to stumble upon the missing element, which they eventually called promethium.

Element 61 in the periodic table represents a curious case in the same way that technetium does. It has an atomic number that is not especially high and yet the discovery of the element took until 1945 to be realized. This is because the element is unusually unstable and in fact the only one of the fourteen lanthanoids ranging from La (57) to Yb (70) that is radioactive. The element is frequently described as being so unstable that it does not occur naturally on Earth, but only on some stars. Or at least this is the standard account found in many books and Internet sources. The full story is inevitably far more complicated.

In fact, promethium does occur naturally on Earth, in extremely miniscule amounts in the mineral apatite as first reported in 1965 by Erämetsä,[19] followed by further reports by Kuroda, who also found traces of the element in pitchblende in 1968. The amount reported by Kuroda and colleagues was $(4 \pm 1) \times 10^{-15}$ grams of ^{147}Pm per kilogram of pitchblende.[20]

Nuclear Batteries

Many isotopes have been used to make nuclear batteries, meaning batteries that create electrical power directly from radioactive decay. They include isotopes like tritium, or ^{3}H, and strontium in the form of ^{90}Sr. While tritium is a low b emitter, ^{90}Sr stands close to the other end of the scale. Meanwhile, ^{147}Pm represents something of an ideal compromise, being a medium β emitter while also not producing too much undesirable secondary radiation while in operation. Not surprisingly, much attention is being focused on the development of ^{147}Pm-based nuclear batteries.[21]

Nuclear batteries in general are expensive but tend to have very long half-lives, typically of ten to twenty years, when compared with conventional chemically based batteries. Nuclear batteries are therefore excellent for applications such as spacecraft, but have also been used to power hearing aids and heart pacemakers where it is desirable not to change one's power source too frequently. Comparing the relative power output of a chemical battery such as lithium and a ^{210}Po nuclear battery shows a staggering one to ten thousand ratio in favor of the nuclear battery.

Returning to ^{147}Pm, this isotope decays specifically via β decay and provides a high energy power source with a maximum of 220 keV and a half-life of 2.6 years. Until recently, promethium-based batteries used in space and military applications were quite large, whereas a team from the University of Missouri has produced batteries the size of a US penny and are aiming to reduce the thickness of such batteries further to the thickness of a human hair. Such batteries hold up to a million times the charge of a conventional battery and incorporate a liquid rather than a solid semiconductor to minimize the damage caused by the energy that is typically generated.[22]

Chapter 10

From Missing Elements to Synthetic Elements

The periodic table consists of about ninety elements that occur naturally, ending with element 92, uranium. One or two of the first ninety-two elements are variously reported either as not occurring on Earth or as occurring in miniscule amounts. To add to the complications in drawing a sharp line between natural and synthetic elements, the element technetium was first created artificially and only later found to occur naturally on Earth in minute amounts.

As we have seen in previous chapters, chemists and physicists have succeeded in synthesizing some of the elements that were missing between hydrogen (1) and uranium (92), such as promethium and astatine. But in addition, a further twenty-five or so new elements beyond uranium have been synthesized, although again one or two of these, such as neptunium and plutonium, were later found to exist naturally in exceedingly small amounts.

At the time of writing, the heaviest element for which there is good experimental evidence is element-118. All other elements between 92 and 118 have also been successfully synthesized including element-117, which was announced in April of 2010. The synthesis of this element means that for the first time, and probably the last, every single space in a contemporary periodic table has been filled, although some of these elements are still awaiting official ratification.[1]

The synthesis of any element involves starting with a particular nucleus and subjecting it to bombardment with small particles with the aim of increasing the atomic number and hence changing the identity of the nucleus in question. More recently, the method of synthesis has changed so that two nuclei of considerable weights are made to collide with the aim of forming a larger and heavier nucleus.

In a sense in which all these syntheses are descended from a key experiment, conducted by Rutherford and Soddy in 1919 at the University of Manchester, Rutherford and Soddy bombarded nuclei of nitrogen with α particles (helium ions) with the result that the nitrogen nucleus was transformed into that of another element.

$$^{14}_{7}N + ^{4}_{2}He \rightarrow ^{17}_{8}O + ^{1}_{1}H$$

Although they did not initially realize it, they had produced an isotope of oxygen. Rutherford and his colleague had achieved the first-ever transmutation of one element into a completely different one.[2] The dream of the ancient alchemists of transforming one element into another one had become a reality, a breakthrough that has continued to yield new elements right up to the present time.

In fact, Rutherford and Soddy did not produce a completely new element but just an unusual isotope of an existing element. They used α particles produced by the radioactive decay of other unstable nuclei such as uranium. Similar transmutations could be carried out with target atoms other than nitrogen but extending only as far as calcium, which has an atomic number of 20. If heavier nuclei were to be transmuted they would require more energetic projectiles than naturally produced α particles and such more energetic projectiles did not become available for a while.

The situation changed in the 1930s, when Ernest Lawrence invented the cyclotron at the University of California, Berkeley. This machine made it possible to accelerate α particles to hundreds and even thousands of times the speed of α particles produced by natural decay processes.[3] A little later another projectile particle, the neutron, was discovered in 1932, with the further advantage of a zero electric charge—which meant that it could penetrate a target atom without suffering any repulsion from the positively charged protons inside the nucleus.[4]

Missing Elements and Transuranium Elements

In the mid-1930s, four gaps remained to be filled in the then existing periodic table. They consisted of the elements with atomic numbers of 43, 61, 85, and 87 as described in previous chapters. Interestingly, the existence of three of these elements had been clearly predicted by Mendeleev many years before and called by him eka-manganese (43), eka-iodine (85), and eka-caesium (87). As we saw, three of these four missing elements—technetium, astatine, and promethium—were first discovered as a result of their being artificially synthesized in the twentieth century.

We come now to the synthesis of the elements beyond the original 1–92. In 1934, three years before the synthesis of technetium, Enrico Fermi, working in Rome, began bombarding element targets with neutrons in an attempt to synthesize some transuranium elements. Fermi soon believed that he had succeeded in producing two such elements, which he named ausonium (93) and hesperium (94).[5] But it was not to be and these elements must be added to the long list of spurious elements that never materialized.

After announcing the findings at his Nobel Prize acceptance speech of the same year, Fermi quickly retracted the claim in the written version of his Nobel presentation. The explanation for Fermi's erroneous claim emerged one year later, in 1938, when Hahn, Strassmann, and Meitner discovered nuclear fission. It then became clear that, on collision with a neutron, the uranium nucleus broke up to form two middle-sized nuclei rather than transforming itself into a larger one as Fermi had thought. For example, uranium formed cesium and rubidium by the following fission reaction, which was first described by Hahn, Strassmann, and Meitner.

$$^{235}_{92}U + ^{1}_{0}n \rightarrow ^{144}_{55}Cs + ^{90}_{37}Rb + 2^{0}_{1}n$$

Fermi and his collaborators had in fact been observing such products of nuclear fission processes instead of forming heavier nuclei as they first believed.

Real Transuranium Elements

The synthesis and identification of the first genuine transuranium element, element 93, was finally carried out in 1939 by Edwin McMillan and his collaborators at the University of California, Berkeley.[6] It was named neptunium because it followed uranium in the periodic table just as the planet Neptune follows Uranus as one moves away from the sun. Philip Abelson, a chemist on the team, discovered that element 93 did not behave as eka-rhenium as expected from its presumed position in the periodic table. On the basis of this and similar findings on element 94, or plutonium, another Berkeley chemist, Glenn Seaborg, proposed a major modification to the periodic table (fig. 3.3). As a result, the elements from actinium (89) onward would no longer be regarded as transition metals but as analogues of the lanthanide series. Consequently, there was no need for elements such as 93 and 94 to behave like eka-rhenium and eka-osmium, since they had been assigned to different places on the revised periodic table.

Elements 94 to 100

The synthesis of elements 94 to 97 inclusive, named plutonium, americium, curium, and berkelium took place in the remaining years of the 1940s while number 98, or californium, was announced in 1950. This sequence looked as though it was about to end since the heavier was the nucleus, the more unstable it became. While it was necessary to accumulate enough target material in order to bombard it with neutrons, to transform the element into a heavier one, such accumulation was becoming increasingly difficult.

At this point, serendipity intervened when a thermonuclear test explosion, code name Mike, was carried out close to the Marshall Islands in the Pacific Ocean in 1952. One of the outcomes of this event was the production of intense streams of neutrons that enabled nuclear reactions that would not otherwise have been possible. For example, the U-238 isotope collides with as many as 15 neutrons to form U-253, which subsequently undergoes the loss of seven β particles, resulting in the formation of element 99, named einsteinium.

$$^{238}_{92}U + 15\,^{1}_{0}n \rightarrow {}^{235}_{92}U \rightarrow {}^{253}_{99}Es + 7\,^{0}_{-1}\beta$$

Element 100, named fermium, was created in a similar manner, as a result of the high neutron flux produced by the same explosion. The presence of element 100 was revealed by analysis of the soil from some nearby Pacific Islands.[7]

From 101 to 106

Moving further along the sequence of heavier nuclei required a quite different approach because β decay does not take place for elements above $Z = 100$. Several technological innovations were required, including the use of linear accelerators rather than utilizing cyclotrons. With linear accelerators, highly intense beams of ions can be produced at well-defined energies. Moreover, projectile particles heavier than neutrons or α particles could now also be used. During the Cold War period, the only countries that possessed such facilities were the two superpowers, the United States and the Soviet Union.

In 1955 mendelevium, element 101, was produced in this way at the linear accelerator at Berkeley:

$$^{4}_{2}He + ^{253}_{99}Es \rightarrow ^{256}_{101}Md + ^{1}_{0}n$$

The American scientists involved in this discovery named the new element after the Russian Mendeleev, in a positive moment in an otherwise tense period between the two nations. The new linear accelerator techniques also rendered possible new combinations of nuclei. For example, element 104, or rutherfordium, was made in Berkeley with the following reaction,[8]

$$^{12}_{6}C + ^{249}_{98}Cf \rightarrow ^{257}_{104}Rf + 4\,^{1}_{0}n$$

In Dubna, Russia, a different isotope of the same element was created in the reaction,

$$^{22}_{10}Ne + ^{242}_{94}Pu \rightarrow ^{259}_{104}Rf + 5\,^{1}_{0}n$$

In all, six elements, from 101 to 106, were synthesized by this approach. As a result of Cold War tension between the United States and Soviet Union, in addition to the usual nationalistic disputes that seem to occur over the discovery of new elements, claims for the synthesis of most of these elements gave rise to controversies that continued for many years to come.

But having reached element 106, a new problem arose that required yet another new approach. At this time, German scientists entered the field with the establishment of the GSI, Institute for Heavy-Ion research in Darmstadt. The new technology was named "cold fusion" although it is not connected with the discredited cold fusion in test tubes that was later announced by the chemists Fleischmann and Pons in 1989.[9]

Cold fusion in the transuranium field is a technique in which nuclei are made to collide with one another at slower speeds than were previously used. As a result, less energy is generated and so there is a decreased possibility that the combined nucleus will disintegrate. The technique was originally devised by a Soviet physicist, Yuri Oganessian, but was perfected in the German facility.

Elements 107 Onward

In the early 1980s elements 107 (bohrium), 108 (hassium), and 109 (meitnerium) were successfully synthesized in Germany using the cold-fusion method[10], until another roadblock became apparent. By this time the Berlin Wall had fallen and the United States and what now became Russia began a fruitful collaboration along with the German team. In 1994, after ten years of stagnation, the German lab in Darmstadt announced the synthesis of element 110 formed by the collision of lead and nickel ions,

$$^{208}_{82}Pb + ^{62}_{28}Ni \rightarrow ^{270}_{110}Ds \rightarrow ^{269}_{110}Ds + ^{1}_{0}n$$

The half-life of element 110, later named darmstadtium, was a mere 170 micro-seconds.[11] The Germans called their element darmstadtium following the trend in the earlier naming of berkelium and dubnium by American and Russian teams, respectively. A month later, the German lab had produced element 111, which later became known as roentgenium after Röntgen, the discoverer of X-rays.[12] February 1996 saw the synthesis of the next element in the sequence, 112, which was officially named copernicium in the year 2010.[13] Two further elements, numbers 114 and 116, have been given names of flevorium and livermorium.

113–118

Since 1997, several claims have been published for the synthesis of elements 113 all the way to element 118, the latest being element-117, synthesized in 2010.[14] The fact that an element with an odd atomic number was the most difficult to produce is not surprising given that nuclei with an odd number of protons are invariably more unstable than those with an even number of protons. This is due to the fact that protons, like electrons, have a spin of one-half or minus one-half and enter into energy orbitals, in pairs with opposite spins. Consequently, even numbers of protons frequently produce total spins of zero and hence more stable nuclei than those with unpaired proton spins as occurs in nuclei with odd numbers of protons such as 115 or 117.

Element 114

The synthesis of element 114 was much anticipated because it had been predicted for some time that it would represent an "island of stability"—that is to say, a portion of the table of nuclei with enhanced stability. This element was first claimed by the Dubna lab in Russia in late 1998 but only definitely produced in further experiments in 1999 involving the collision of a plutonium target with calcium-48 ions.[15] The labs at Berkeley and Darmstadt have recently confirmed this finding. At the time of writing something like 80 decays involving element 114 have been reported, 30 of which come from the decay of heavier nuclei such as 116 and 118. The longest lived isotope of element-114 has a mass of 289 and a half-life of about 2.6 seconds, in agreement with predictions that this element would show enhanced stability, although the greater degree of stability displayed by the element is regarded as something of a disappointment by some experts.[16]

Element 118

On December 30, 1998, the Dubna-Livermore labs published a joint paper, claiming element 118 had been observed as a result of the following reaction,[17]

$$^{86}_{36}Kr + ^{208}_{82}Pb \rightarrow ^{293}_{118}Uno + ^{1}_{0}n$$

After several failed attempts to reproduce this result in Japan, France, and Germany, the claim was officially retracted in July of 2001. The controversy surrounding this case included the dismissal of a senior member of the Berkeley research team who had published the original claim.[18]

A couple of years later, new claims were announced from Dubna and followed in 2006 by further claims by the Lawrence-Livermore Laboratory in California. Collectively, the US and Russian scientists made a stronger statement that they had detected four more decays of element 118 from the following reaction,

$$^{249}_{98}Cf + ^{48}_{20}Ca \rightarrow ^{294}_{118}Uno + 3\,^{1}_{0}n$$

The researchers are now highly confident that these results are reliable, and point out that the chance that the detections were random events is less than one part in 100,000.[19] Needless to say, no chemical experiments have yet been conducted on this element in view of the paucity of atoms produced and their very short lifetimes of less than one millisecond.

In 2010, an even more unstable element, number 117, was synthesized and characterized by a large team of researchers working in Dubna, as well as several labs in the United States. The periodic table has reached an interesting point, at which all 118 elements exist either in nature or have been created artificially in special experiments. This includes a remarkable twenty-six elements beyond the element uranium, which has been traditionally regarded as the last of the naturally occurring elements. Attempts to create yet heavier elements such as 119 and 120 have already begun and there are no reasons for believing that there should be any immediate end to the sequence of elements that can be formed. The question of "the end of the periodic table" has led to much debate and disagreement among experts, although some believe that element number 137 may well mark the upper limit to the production of new elements.[20] On the other hand, more accurate estimates that take account of the finite volume of the nucleus lead to predictions of an upper limit of 172 or 173.[21]

Chemistry of the Synthetic Elements

The existence of superheavy elements raises an interesting new question as well as a challenge to the periodic table. It also affords an intriguing new arena for theoretical predictions to be compared against experimental findings. Theoretical calculations suggest that the effects of relativity become increasingly important as the nuclear charge of atoms increases. For example, the characteristic color of gold, with a rather modest atomic number of 79, is now explained by appeal to relativity theory. The larger the nuclear charge, the faster the motion of inner shell electrons. As a consequence of gaining relativistic speeds, inner electrons are drawn closer to the nucleus and this in turn has the effect of causing greater screening on the outermost electrons, which determine the chemical properties of any particular element. It is predicted that some atoms will behave chemically in a manner that is unexpected from their presumed positions in the periodic table.

Relativistic effects thus pose the latest challenge to test the scope and universality of the periodic table. When elements 104 and 105, rutherfordium and dubnium, respectively, were chemically examined, the situation reached something of a climax. It emerged that the chemical behaviors of rutherfordium and dubnium were in fact rather different from what was expected considering where these elements lie in the periodic table. Rutherfordium and dubnium did not seem to behave like hafnium and tantalum, respectively, as they should have done if the periodic table remains valid up to such high atomic numbers.[22]

For example, in 1990 Czerwinski, working at Berkeley, reported that the solution chemistry of element 104, or rutherfordium, differed from that of zirconium and hafnium, the two elements lying above it. He also reported that rutherfordium's chemistry resembled that of the element plutonium, which lies quite far away in the periodic table. Meanwhile, early studies on dubnium showed that it, too, was not behaving like the element above it, namely tantalum.

Instead, dubnium showed greater similarities with the actinide element of protactinium. In other experiments rutherfordium and dubnium seemed to be behaving more like the two elements above hafnium and tantalum, namely zirconium and niobium (fig. 10.1).

It was only when the chemistry of elements seaborgium (106) and bohrium (107) was examined that they showed that the expected periodic behavior was resumed and that the status of the periodic table remained secure. The titles of the articles that announced these discoveries spoke for themselves. Such headlines as "Oddly Ordinary Seaborgium" and "Boring Bohrium," were published, both references to the fact that it was business as usual for the periodic table.[23] Even though relativistic effects might have been even more pronounced for these two elements, the expected chemical behavior seems to outweigh any such tendencies. Seaborgium and bohrium were behaving as the periodic table demands that they should.

The fact that bohrium behaves as a good member of group 7 can be seen from the following argument that I have proposed in some publications.[24] This approach also represents a "full circle" of sorts, since it involves a triad of elements. As the reader may recall from chapter 1, the discovery of triads was the very first hint of a numerical regularity relating the properties of elements within a common group. Here now is the data for measurements carried out on the standard sublimation enthalpies, of analogous compounds of technetium, rhenium, and bohrium with oxygen and chlorine (energy required to convert a solid directly into a gas) (fig. 10.2).[25]

3	4	5	6	7	8	9	10	11	12
Sc	Ti	V	Cr	Mn	Fe	Co	Ni	Cu	Zn
Y	Zr	Nb	Mo	Tc	Ru	Rh	Pd	Ag	Cd
Lu	Hf	Ta	W	Re	Os	Ir	Pt	Au	Hg
Lr	Rf	Db	Sg	Bh	Hs	Mt	Ds	Rg	Cn

FIGURE 10.1 Fragment periodic table showing groups 3–12 inclusive.

TcO_3Cl = 49 kJ/mol
ReO_3Cl = 66 kJ/mol
BhO_3Cl = 89 kJ/mol

FIGURE 10.2 Sublimation enthalpies of elements in group 6.

Tc
Re
Bh

FIGURE 10.3 Group 7.

Predicting the value for BhO_3Cl using the triad method gives 83 kJ/mol, or an error of only 6.7 percent compared with the above experimental value of 89 kJ/mol (fig. 10.2). This fact would seem to lend further support to the notion that bohrium acts as a genuine group 7 element (fig. 10.3).

The challenge to the periodic law from relativistic effects became even more significant in the case of number 112, or copernicium, one of the most recent elements for which chemical experiments have been conducted.[26] Once again, relativistic calculations indicated modified chemical behavior to the extent that the element was thought to behave like a noble gas rather than like mercury, below which it is placed in the periodic table. Experiments carried out on sublimation enthalpies on element 112 then showed that contrary to earlier expectations, the element truly belongs in group 12 along with zinc, cadmium, and mercury as shown in fig. 10.4.

Element 114 presented a similar story with early calculations and experiments suggesting noble gas behavior but more recent experiments supporting the notion that the element behaves like the metal lead as expected from its position in group 14. The conclusion would seem to be that chemical periodicity is a remarkably robust

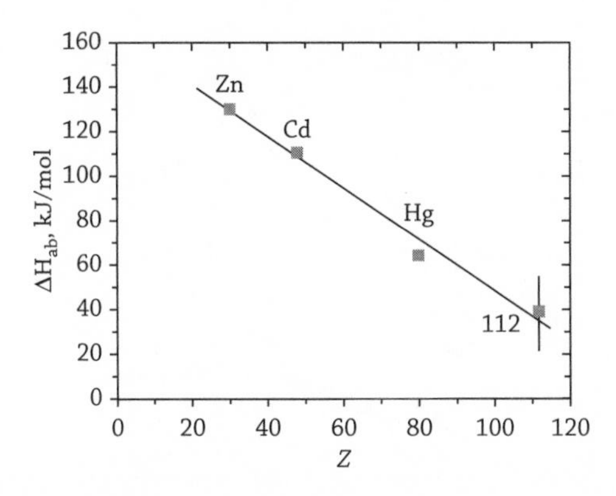

FIGURE 10.4 Sublimation enthalpies for elements in group 12 showing that element 112 is a genuine member of this group. (From Gas Phase Chemistry, H. W. Gäggeler. A. Türler, In *The Chemistry of the Superheavy Elements*, M. Schadel, Kluwer Academic Publisher, Dordrecht, 2003). Reprinted with permission of Springer Science and Business Media.

phenomenon. Not even the powerful relativistic effects due to fast moving electrons seem to be capable of toppling a simple scientific discovery that was made around 150 years ago, but of course one must be open to possible surprises.

Moreover, at least for the foreseeable future it does not look as if relativity theory is about to cause a major upset to the periodic table. This is because relativistic effects do not simply increase with increasing atomic number. It is more a case of the interplay of relativistic effects and quantum effects that govern the particular order of energy levels in any given atom. For example, in the sixth period, the largest relativistic effect occurs at the atoms of gold and not at the end of the period where the atomic number is even higher (fig. 10.5). This effect has been termed the gold maximum phenomenon by the Finnish chemist and leading expert in relativistic quantum chemistry, Pekka Pyykkö. It is because the element gold shows the largest relativistic effect of any element in its period that causes it to display anomalous chemical and physical properties. For example, gold displays a characteristic golden-yellow color, unlike any of its surrounding transition

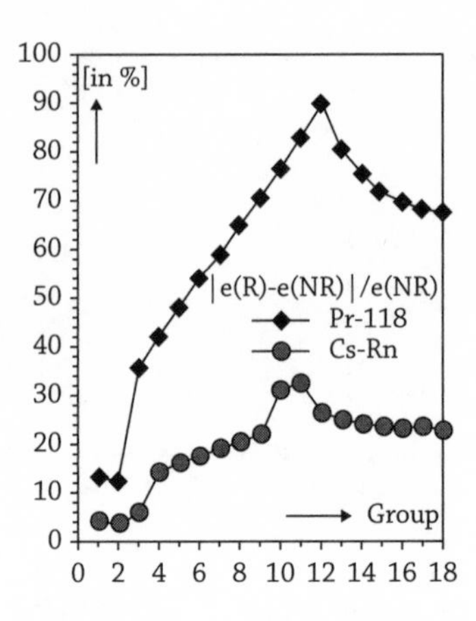

FIGURE 10.5 Diagram of gold maximum effect (circles) as well as the analogous case for the subsequent period at element 112 (diamonds). Reprinted with the permission of John Wiley and Sons.

From P. Schwerdtfeger, M. Seth, Relativistic Effects of the Superheavy Elements, in *The Encyclopedia of Computational Chemistry* eds P. v. R. Schleyer, N. L. Allinger, T. Clark, J. Gasteiger, H. F. Schaefer III, P. R. Schreiner, John Wiley and Sons, New York, 1998.

metals in the periodic table. It also shows an anomalous voltage when used to make up an electrical cell and has a tendency to form unusual oxidation states and a host of unexpected new compounds, many of which were predicted by Pyykkö.[27]

Calculations carried out by others have shown that in period seven the maximum relativistic effect should take place at element 112.[28] As can be seen from fig. 10.3, the effect then drops away rather sharply for subsequent elements. The fact that the chemistry of elements 112 and 114 has been examined and that results show no signs of highly anomalous behavior would seem to suggest that the chemistry of the subsequent elements should also behave as expected on the basis of the periodic table. This seems to be a further testament to the underlying fundamental nature of the periodic law, which continues to stand firm against the threats from quantum mechanics and relativity combined together.

NOTES

Introduction

1. S. Kean, *The Disappearing Spoon*, Little, Brown and Co., New York, 2010; H. Aldersey-Williams, *Periodic Tales*, Penguin Viking, London, 2011.

2. Websites: http://www.meta-synthesis.com/webbook/35_pt/pt_database.php?PT_id=406 http://guitarnoize.com/the-periodic-table-of-guitarists/

3. J. Emsley, *The Elements*, 3rd Edition, Clarendon Press, Oxford, 1998.

4. Eric, R. Scerri, *The Periodic Table, Its Story and Its Significance*, Oxford University Press, New York, 2007.

5. V. Karpenko, The Discovery of Supposed New Elements, *Ambix*, 27, 77–102, 1980.

6. F. Habashi, The History of Element 43—Technetium, *Journal of Chemical Education*, *83*, 213–213, 2006.

7. If we also incorporate theoretical developments concerned with the periodic table, we would need to include Maria Goeppert-Meyer, Charlotte Moore, Birtha Jeffreys, and several other women.

8. This is only true in principle. In fact, Moseley could only state that there were four missing elements with any confidence.

9. Another way to identify these elements would be to say that they were the seven missing elements between the limits of the "old periodic table" consisting of elements 1–92.

10. The most recently synthesized element was number 117, which was announced in 2010. This means that for the first time, and perhaps for the last time, the periodic table is absolutely complete with not one single gap along any of the rows. This feature will disappear just as soon as element 119 is synthesized since this will signal the start of an extremely long period of fifty elements, most of which will never even see the light of day.

11. S. Lyle, Narrative understanding: developing a theoretical context for understanding how children make meaning in classroom settings, *Journal of Curriculum Studies*, 32, 45–63, 2000.

12. J. Emsley, *The A-Z of the Elements*, Oxford University Press, Oxford, 2001; A. Swertka, *A Guide to the Elements*, Oxford University Press, New York, 1998.

13. Letter from Meitner to Hahn, February 22, 1911.

14. Strictly speaking, the term infra-uranium would have to exclude uranium itself.

15. The periodic table was independently discovered by at least six scientists of whom Mendeleev was the last, although admittedly and by far the most successful. See my book on the periodic table. E. R. Scerri, *The Periodic Table, Its Story and Its Significance*, Oxford University Press, New York, 2007.

16. J. Levy, *Scientific Feuds*, New Holland, London, 2010.

17. R. Merton, Priorities in Scientific Discovery, *American Sociological Review*, 635–659, 1957.

18. For example, it was Wollaston's friends, rather than Wollaston himself, who insinuated that the young Faraday had taken credit for the experiments on electromagnetic rotation.

19. Of course, it would have been rather odd if a sociologist would have concluded otherwise.

20. The dispute was not settled until 1997, at which time members of the IUPAC committee ruled in favor of the American claim.

21. This work includes A. Brannigan, *The Social Basis of Scientific Discoveries*, Cambridge University Press, Cambridge, 1981; G. Markus, Why Is There No Hermeneutics of Natural Sciences? Some Preliminary Theses, *Science in Context*, *1*, 5–15, 1987; M. Mulkay, Norms and Ideology in Science, *Social Science Information*, *15*, 637–656, 1976.

22. A. G. Gross, Do Disputes over Priority Tell Us Anything about Science? *Science in Context*, *11*, 161–179, 1998.

23. E. Rancke-Madsen, The Discovery of an Element, *Centaurus, 19,* 299–313, 1976.

24. I have made the same point about the discovery of the periodic table in my earlier book on the subject.

25. Glenn Seaborg first announced the discovery of plutonium, element 94, on a children's radio quiz show.

26. T. S. Kuhn, *Historical Structure of Scientific Revolutions,* 2nd ed., University of Chicago Press, Chicago, 1970, p. 55.

27. Roald Hoffmann and Carl Djerassi, *Oxygen,* a play, Wiley-VCH, 2001.

28. There is an interesting parallel here with the question of whether Mendeleev should be given the major credit for the discovery of the periodic table even though he was not the first discoverer. The widely held view is that he should, since he made more of the discovery than his competitors.

29. Before Lavoisier's time many of what we would term compounds were regarded as being elementary while our elements were believed to be composite bodies.

30. T. S. Kuhn, Historical Structure of Scientific Discovery, *Science, 136,* 760–764, 1962.

31. I give their atomic numbers rather than their modern names in order not to prejudge the issue of their discovery.

32. A. F. Holleman, E. Wiberg, *Inorganic Chemistry,* Academic Press, San Diego, CA, 2001, p. 1695.

33. Eric R. Scerri, *The Periodic Table, Its Story and Its Significance,* Oxford University Press, New York, 2007. I have also taken the opportunity to correct any mistakes that appeared in the 2007 book in these two chapters, while starting to plan a revised edition.

Chapter 1

1. P. Needham, Has Daltonian Atomism Provided Chemistry with Any Explanations? *Philosophy of Science, 71,* 1038–1047, 2004; P. Needham, When did atoms begin to do any explanatory work in chemistry? *International Studies in the Philosophy of Science, 18,* 199–219, 2004.

2. P. J. Hartog, A First Foreshadowing of the Periodic Law, *Nature, 41,* 186–188, 1889; P. E. Lecoq De Boisbaudran, A. Lapparent, A Reclamation of Priority on Behalf of M. De Chancourtois Referring to the Numerical Relations Among Atomic Weights, *Chemical News, 63,* 51–52, 1891.

3. This name was also chosen by reference to tellos, the Greek for Earth, by De Chancourtois, who was a geologist and so is primarily interested in classifying the elements of the earth.

4. Among the recipients of De Chancourtois's privately published system was Prince Napoleon.

5. A. E. Béguyer De Chancourtois, *Compes Rendus de l'Académie des Sciences, 54*, 1862, 757–761, 840–843, 967–971.

6. These failures can be attributed to the existence of the lanthanide elements, which occur between the second and third transition series of elements in modern terms. The lanthanides would be a problem for all the discoverers of the periodic system, as only six of the fourteen of these elements had been discovered prior to the 1860s, when these early periodic systems were being developed.

7. A very similar table comparing differences in atomic weights between first and second members of groups of analogous elements is discovered independently and published in the very same year by Odling, as will be seen below. Indeed, Odling outdoes Newlands in recognizing ten such relationships, to Newlands's six.

8. A periodicity of eight was correct for the chemistry known at the time. Today the periodicity is actually nine, counting from the first element up to and including the first analogous element, (e.g., from lithium to sodium), as discussed in chapter 1.

9. J. A. R. Newlands, On the Law of Octaves, *Chemical News, 12,* August 18, 83–83, 1865.

10. J. A. R. Newlands, On the Law of Octaves, *Chemical News, 13,* 130–130, 1866.

11. On the other hand, Newlands can be faulted for omitting gaps for as yet undiscovered elements in the manner that Mendeleev later included.

12. Here I am meaning the distance between the number of successive similar elements to be consistent with the Newlands quotation and not as in other parts of the present book when considering one element up to and including its analogue.

13. W. Odling, On the Proportional Numbers of the Elements, *Quarterly Journal of Science, 1,* 642–648, October 1864.

14. Ibid.

15. Van Spronsen correctly praises Odling, in my view, for being the first to recognize this feature, although I differ somewhat regarding the details, as argued in the main text.

16. Carl A. Zapffe, Hinrichs, Precursor of Mendeleev, *Isis*, *60*, 461–476, 1969. Quotation from p. 464.

17. The connection is altogether different from that postulated by Hinrichs, however.

18. Clearly, the correspondence with the astronomical distances is only approximate.

19. Isaac Newton is credited with first performing a similar experiment with sunlight, which he dispersed into its component colors, also by means of a glass prism.

20. It is said that Bunsen never once referred to the work of his former students Mendeleev and Lothar Meyer, either in writings or in the course of lectures. This was in spite of the fact that both of these former students acquired considerable fame for their respective systems of classifying the elements.

21. According to the atomic weights used by Hinrichs, calcium has a weight of 20 and barium a weight of 68.5.

22. In the modern table these three metals are indeed grouped together, but not in the same group as carbon and silicon, which belong with germanium, tin, and lead in group 4.

23. For example, Hinrichs groups together O, S, Se, and Te. Newlands also groups these elements together but includes osmium (Os) with them. Hinrichs groups together N, P, As, Sb, and Bi. So does Newlands, but he incorrectly includes Mn, as well as Di and Mo in one space. Hinrichs groups together Li, Na, K, and Rb. Newlands also groups these elements together but also incorrectly includes Cu, Ag, Cs, Au, and Tl.

24. G. Hinrichs, *The Elements of Chemistry and Mineralogy*, Griggs, Watson & Day, Davenport, Iowa, 1871; G. Hinrichs, *The Principles of Chemistry and Molecular Mechanics*, Day, Egbert & Fidlar, Davenport, Iowa, 1874.

25. This is not to say that his classification is unconnected with atomic weights, only that the connection is rather indirect in view of the astronomical argument that seems to be the basis of the approach.

26. However, his spectral studies are by no means universally accepted. Some contemporary historians, including Klaus Hentschel, have criticized Hinrichs's work, claiming that he was somewhat selective in what data he admitted into his calculations.

27. Part of the motivation for Cannizzaro's work on atomic weights lies with the earlier work of Avogadro as mentioned previously.

28. Lothar Meyer in his editorial on the papers of Cannizzaro in Oswald's *Klassiker der Exacten Wissenschaften*: 30, Arbis eines Lehrganges der theoretischen Chemie, vorgetragen von Prof. S. Cannizzaro, Leipzig, 1891.

29. It should perhaps be noted in passing that it took Mendeleev something like nine years from the time of his attending the same conference before he, too, produced a table of elements arranged in order of increasing atomic weights.

30. The term horizontal relationship may be a little ambiguous given that some tables show chemical groups vertically and others horizontally. I am using the term here in the sense previously mentioned in connection with Lenssen to mean relationships between elements that are not chemically analogous, or elements with steadily increasing atomic weights. These relationships appear horizontally as periods in the modern table and indeed in many but not all tables of the Lothar Meyer–Mendeleev period.

31. In the modern table one sees an initial increase from one to four followed by a decrease down to one again once the halogens are reached. Lothar Meyer's table differs from the modern one simply in that he chooses to begin with the modern group 4. In addition, the noble gases had not yet been discovered in 1864 and the modern group 3 had not yet been recognized as a separate group.

32. As was mentioned in the case of Odling, such a separation has become a feature of the modern medium-long form and long form tables.

33. The criticism has been made that Lothar Meyer did not explicitly refer to atomic weight in his 1864 table. This objection seems a little excessive, however, since the twenty-eight-element arrangement is clearly based on increasing atomic weight, such that Lothar Meyer may not feel the need to comment on this rather obvious feature. Of course, the same cannot be said for the smaller table consisting of twenty-two elements. But the fact that these elements have been separated from the other twenty-eight may indicate that Lothar Meyer realized that in these cases the concept of increasing atomic weight did not apply strictly to the classification he chose to adopt. Nevertheless, atomic weight increases vertically down each column, and there are only six inconsistencies in the increase in atomic weight going across the table. Given that Lothar Meyer had classified a total of fifty elements while only showing six mistaken reversals in atomic weights, all of which occur among the problematic transition metals (in the modern usage of the term), this cannot be considered a significant failing on his part. Indeed, the only serious misplacements he made in terms of atomic

weight increase concern just two elements, molybdenum and vanadium. All of his other reversals are quite within the possible bounds of error in measured atomic weights.

34. In his famous table of 1869 Mendeleev wrongly placed mercury with copper and silver, misplaced lead with calcium, strontium, and barium, and also misplaced thallium among the alkali metals. For a more detailed set of comparisons, see van Spronsen's book, p. 127–131. The misplacement of mercury with silver is perhaps not altogether surprising given that hydrargyrum, the Latin name for mercury, means "liquid silver."

35. J. E. Earley, How chemistry shifts horizons: element, substance, and the essential, *Foundations of Chemistry, 11,* 65–77, 2009; R. F. Hendry, *Lavoisier and Mendeleev on the Elements, Foundations of Chemistry,* 7, 31–48, 2005; E. R. Scerri, What is an element? What is the periodic table? And what does quantum mechanics contribute to the question? *Foundations of Chemistry, 14,* 69–81, 2012.

36. In previous publications I claimed otherwise. I thank Philip Stewart for setting me straight on this point.

37. Although this very early table shows a value of 5.5 for oxygen, Dalton soon changed this to 8.0, the value discussed in the text here.

Chapter 2

1. A. Einstein, *Investigations of the Theory of the Brownian Movement,* with notes by R. Fürth, translated by A. D. Cowper, Dover Publications, Mineola, New York, 1956.

2. J. Perrin, Mouvement brownien et réalité moléculaire, *Annales de Chimie et de Physique, 18,* 1–114, 1909.

3. This is another notion that Mendeleev had strenuously opposed during his lifetime.

4. His Cambridge predecessor, Thomson, had thought that the atom consisted of a sphere of positive charge within which the electrons circulated in the form of rings.

5. A. J. van den Broek, The α Particle and the Periodic System of the Elements, *Annalen der Physik,* 23, 199–203, 1907.

6. A. J. van den Broek, The Number of Possible Elements and Mendeléeff 's "Cubic" Periodic System, *Nature,* 87, 78, 1911.

7. N. Bohr, On the Constitution of Atoms and Molecules, *Philosophical Magazine, 26,* 1–25, 476–502, 857–875, 1913 (known as the trilogy paper). Van den Broek is cited on p. 14.

8. H. G. J. Moseley, Atomic Models and X-Ray Spectra, *Nature, 92,* 554, 1913.

9. The Xs denoted unknown species that turned out to be isotopes of different elements in most cases. For example, uranium X was later recognized to be an isotope of thorium.

10. It was also not wasted work because Paneth and von Hevesy's efforts allowed them to establish a new technique for the radioactive labeling of molecules, which became the basis for a very useful subdiscipline with far-reaching applications in areas like biochemistry and medical research.

11. Nevertheless, Einstein soon began to regard quantum mechanics as an incomplete theory and was to maintain his criticism of it for the remainder of his life.

12. Both of these assumptions are somewhat *ad hoc*. For example, Bohr did not deduce why the electrons occurred in quantized stationary states, but merely asserted that they did.

13. The overwhelming majority of chemistry and physics textbooks claim that this Madelung, or n + ℓ, rule provides the order of filling of atomic orbitals. As a matter of fact, as recently pointed out by Schwarz, it only provides the order of orbital filling for the s-block elements and none of the elements in the remaining three blocks of the periodic table. Nevertheless, the rule is still important since it continues to provide the overall configuration of most atoms in the periodic table, if not the precise order in which each orbital is occupied. W. H. E. Schwarz, *Journal of Chemical Education*, 87, 444–448, 2010. Also, see my recent blog in which I call the usually presented version the "sloppy aufbau," http://ericscerri.blogspot.com/

14. In my view, many philosophers of science have been carried away with the notion that reduction among the sciences is doomed to failure. Fortunately, some scientists, such as physicist Steven Weinberg, have reminded them otherwise. Weinberg, Steven, "Reductionism Redux," *The New York Review of Books*, October 5, 1995. Reprinted in Weinberg, S., *Facing Up*, Harvard University Press, 2001.

Chapter 3

1. Mendeleev, as cited in Smith, J. R. *Persistence and Periodicity*, unpublished PhD thesis, University of London. 1975, p. 401.

2. In terms of atomic weights iodine should be placed before tellurium in the periodic table but this would imply that tellurium is a halogen and that iodine is a member of the oxygen group. In fact, the opposite is true in chemical terms. The early discoverers of the periodic system overcame this problem by making the reversal and overriding the order of atomic weight increase. Mendeleev believed that the atomic weights of either iodine or tellurium had been incorrectly determined and challenged the experimentalists to repeat their measurements. In spite of many such attempts, the atomic weight ordering did not change. Iodine does indeed have a lower atomic weight than does tellurium.

3. Not that this fact can be held against Mendeleev's predictive abilities since he made ample use of horizontal trends in many other cases.

4. Crookes was the founder and editor of the journal *Chemical News*. For a biography of Crookes, see William H. Brock, *William Crookes (1832–1919) and the Commercialization of Science*, Ashgate, Aldershot, UK, 2008.

5. Other elements have been named after astronomical bodies including cerium after ceres, palladium after pallas, and more recently, neptunium after Neptune and plutonium after Pluto.

6. It is by no means clear that Becquerel was the first to discover radioactivity contrary to the most accounts and indeed the one given here. See T. Rothman's book, *Everything's Relative*, Wiley, Hoboken, NJ, 2003, p. 46–52. Rothman makes a very good case for the prior discovery by Abel Niépce de Saint-Victor, who was incidentally the brother of Joseph-Nicéphore Niépce, originator of the first ever photographic image.

7. William Crookes, Radio-Activity of Uranium, *Proceedings of the Royal Society of London, 66*, 409–423, 1899–1900.

8. P. Thyssen, *Accommodating the Rare-Earths in the Periodic Table*, M.Sc. Thesis, Catholic University of Leuven, 2009.

9. As quoted by Sime in her book, R. Sime, *Lise Meitner, A Life in Physics*, University of California Press, Berkeley, CA, 1996.

10. Within a couple of years van den Broek's hypothesis, that atomic charge was the defining criterion of any element, was to receive experimental support from Moseley's classic work but this was still in the future.

11. With the exception of radium, Glenn Seaborg would eventually move all three of these elements as well as others to a different part of the periodic table. They came to form part of the actinide series and it was realized that any apparent analogy with the third transition metal series was only partial.

12. Letter from Meitner to Hahn. October 25, 1916. This and subsequently cited letters are from the *Otto Hahn Nachlass* in the Bihliothek und Archiv der Max-Planck- Gesselschaften, Berlin-Dahlem.

13. Letter from Meitner to Hahn, February 22, 1917.

14. Letter from Meitner to Hahn, November 28, 1917.

15. O. Hahn, L. Meitner, Die Muttersubstanz des Actiniums, ein Neues Radioaktives Element von Langer Lebensdauer. *Physikalische Zeitschrift, 19,* 208–218, 1918.

16. Letter from Stefan Meyer to Meitner, 5th June, 1918, (Churchill College Archives, Cambridge).

17. It is very odd to see that many popular books and websites on the discovery of the elements insist on listing Soddy and Cranston as codiscoverers of element 91 and in some cases as the sole discoverers.

18. Hahn was also obliged to erase all mention of Meitner when giving his research presentations.

19. O. Hahn and F. Strassmann, *Die Naturwissenschaften* 27, p. 11–15 (January 1939) received December 22, 1938.

20. Sharon McGrayne, *Nobel Prize Women in Science,* John Henry Press, Washington, D.C., 2002. My account of Meitner and nuclear fission draws heavily from this source.

21. Having now studied this issue more thoroughly, I retract the comment I made in my first book on the periodic table, *The Periodic Table, Its Story and Its Significance,* Oxford University Press, New York & Oxford, 2007.

22. Recall that Mendeleev had predicted the element and had named it eka-tantalum.

23. P. Armbruster, F. P. Hessberger, Making New Elements, *Scientific American.* Sept. 1998, 72–77. This article is followed by one on the history of the periodic system, p. 78–83 by the present author.

24. W. B., Jensen, Classification, Symmetry and the Periodic Table, *Computation and Mathematics with Applications, 12B,* 487–509, 1986; H. Merz, K. Ulmer, Positions of lanthanum and lutetium in the Periodic Table, *Physics Letters,* 26A, 6, 1967; D. C. Hamilton, M. A. Jensen, Mechanism for Superconductivity in lanthanum and uranium, *Physical Review Letters,* 11, 205, 1963; D. C. Hamilton, Position of lanthanum in the periodic table, *American Journal of Physics,* 33, 637, 1965; E.R. Scerri, Which Element Belongs in Group 3?, *Journal of Chemical Education,* 86, 1188–1188, 2009.

25. A. V. Grosse, The Analytical Chemistry of Element 91, *Journal of the American Chemical Society*, 52, 1742–1747, 1930.

26. An interesting new article that brings the chemistry of protactinium right up to date is R. E. Wilson, Peculiar Protactinium, *Nature Chemistry*, 4, 586–586, 2012.

27. Thorium was discovered by the famous Swedish chemist Jöns Jakob Berzelius in 1828. He named thorium after Thor, the Norse god of thunder. The radioactivity of the elements was independently discovered in 1898 by Madame Curie and by Gerhard Carl Shmidt. The discovery of uranium was made before that of thorium in 1789 by the German chemist Martin Heinrich Klaproth working in Berlin.

Chapter 4

1. E. R. Scerri, *The Periodic Table, Its Story and Its Significance*, Oxford University Press, New York, 2007.

2. The history of this field is a complicated one that has essentially involved the multiplication of such elements from there being just four of them in the time of Mendeleev, to the eventual discovery of a further twenty-four to make a total of twenty-eight rare earth elements.

3. The priority dispute between Urbain and von Welsbach has been discussed by Helge Kragh among others, H. Kragh, in Elements No. 70, 71, and 72: Discoveries and Controversies, in C. H. Evans (Ed.), *Episodes in the History of Rare Earth Elements*, 67–89, Kluwer Academic Publishers, 1996.

4. The choice of the name Celtium derives from the Celtic peoples, whose languages survive in parts of Wales, Ireland, and, most importantly for Urbain perhaps, in the French region of Brittany.

5. An associate of Maurice De Broglie, brother of the more famous founder of wave mechanics Louis De Broglie.

6. For a fuller version of this episode, see, E. R. Scerri, *The Periodic Table, Its Story and Its Significance*, Oxford University Press, New York, 2007.

7. As a matter of fact, they first proposed the name danium after Denmark. See H. Kragh, P. Robertson, On the Discovery of Element 72, *Journal of Chemical Education*, 56, 456–459, 1979.

8. D. Coster and G. Hevesy, On the New Element Hafnium, *Nature*, *111*, 462–463, 1923.

9. Ibid.

10. Maurice was the brother of the more famous Louis de Broglie, the discoverer of wave mechanics.

11. Holland, where Coster came from, had also been neutral during the Great War and so could as such be regarded as being on the "wrong side." Hungary had been part of the Austro-Hungarian Empire and so was also regarded with suspicion.

12. Cited in G. Hevesy, *Adventures in Radioisotope Research*, Pergamon Press, Oxford, 1962, p. 11.

13. Another reason for his choosing the name "oceanium" was as a mythological analogue of the element titanium, which lies two places above element 72 in the periodic table.

14. Letter from Scott to Coster and Hevesy, February 5, 1923 (*Bohr Scientific Correspondence, Niels Bohr Institute, Copenhagen*).

15. *The Times* of London, February 2, 1923.

16. H. M. Hansen, S. Werner, Optical Spectrum of Hafnium, *Nature, 111*, 322–322, 1923.

17. G. Urbain, Sur le celtium, élément de numéro atomique 72, *Comptes Rendus de l'Académie des Sciences, 176*, 469–470, 1923.

18. N. Bohr, Letter to Rutherford, February 9, 1923, *Niels Bohr Scientific Correspondence*, Bohr Institute, Copenhagen.

19. N. Bohr, Letter to Rutherford, February 9, 1923, *Niels Bohr Scientific Correspondence*, Bohr Institute, Copenhagen.

20. Rutherford, Letter to N. Bohr, February 15, 1923, *Niels Bohr Scientific Correspondence*, Bohr Institute, Copenhagen.

21. Eric Scerri, Collected Papers in Philosophy of Chemistry, *Imperial College Press*, London, 2008.

22. K. R. Popper, *The Open Universe*, London, Hutchinson, ed. W. W. Bartley III., 1982, pp. 163–164.

23. I have been personally involved in the development of the philosophy of chemistry and in fact had the unique opportunity to meet Karl Popper and discuss this issue with him in London, just a couple of years before he passed away in 1994 at the age of ninety-two. Popper readily accepted some aspects of my critique.

24. Eric Scerri, *The Periodic Table, Its Story and Its Significance*, Oxford University Press, New York, 2007.

25. H. Kragh, Conceptual Changes in Chemistry: The Notion of a Chemical Element, ca. 1900–1925, *Studies in History and Philosophy of Modern Physics, 31*, 435–450, 2000.

26. In addition, one cannot discount an element of nationalism even here. It is apparently perfectly acceptable for a Dane like Kragh to criticize a fellow Dane such as Bohr but not for a foreigner like myself to do so.

Chapter 5

1. Noddack, W., Tacke, I., Berg, O. Die Ekamangane, *Naturwissenschaften 13* (26): 567–574, 1925.

2. Today, rhenium is extracted far more efficiently as the byproduct of the purification of molybdenum and copper.

3. van Assche, P. H. M. The Ignored Discovery of Element Z = 43, *Nuclear Physics, A480,* 205–214, 1988.

4. Kuroda, P. K. A Note on the Discovery of Technetium, *Nuclear Physics, A503,* 178–182, 1989.

5. Yoshihara, H. K. Discovery of a new element 'nipponium': re-evaluation of pioneering works of Masataka Ogawa and his son Eijiro Ogawa, *Spectrochimica Acta Part B Atomic Spectroscopy 59,* 1305–1310, 2004.

6. The current value is 4889 Å and so well within Ogawa's error margin.

7. This was incorrect since a far more common valence for the element, under the conditions that Ogawa was operating with, is hexavalency. If Ogawa had used hexavalency, he would have calculated an atomic weight of 185.2, which is in rather good agreement with the current atomic weight of rhenium (186.2).

8. Rhenium is indeed found in molybdenite, or molybdenum sulfate. This happens because of the close similarities between the ionic radii of molybdenum and rhenium, which means that the latter can replace the former in the crystal lattice. The two elements provide a little-known example of a diagonal behavior, meaning two elements that are very similar by virtue of their diagonal relationship in the periodic table rather than because they lie in the same group. The better known examples of diagonal relationships include lithium and magnesium, beryllium and aluminum, as well as boron and silicon. See G. Rayner-Canhan, Isodiagonality in the Periodic Table, *Foundations of Chemistry, 13,* 121–129, 2011.

9. Yoshihara, H. K. Discovery of a new element 'nipponium': re-evaluation of pioneering works of Masataka Ogawa and his son Eijiro Ogawa, *Spectrochimica Acta Part B Atomic Spectroscopy 59,* 1305–1310, 2004.

10. The story of technetium, which is filled with further controversies, will be the subject of chapter 6.

11. Korzhinsky, M. A., Tkachenko, S. I., Shmulovich, K. I., Taran Y. A., Steinberg, Discovery of a pure rhenium mineral at Kudriavy volcano, G. S. *Nature 369*: 51–52, 2004.

12. Cotton, F. A., Walton, R. A. *Multiple Bonds Between Metal Atoms,* Oxford University Press, Oxford, 1993.

13. Qin J. et al. Is Rhenium Diboride a Superhard Material? *Advances in Materials, 20,* 4780–4783, 2008.

14. S. J. La Placa, B. Post, The Crystal Structure of Rhenium Diboride, *Acta Crystallographica, 15,* 97–99, 1962.

15. This is not to say that there was no criticism of the UCLA research. Objections came mainly from two sources, but all the objections raised have been put to rest in subsequent publications. *Advanced Matererials.,* 21, 4284–4285, 2009; *Adv. Funct. Mater.,* 19, 3519, 2009 and *Acta Materialia,* 58, 1530–1535, 2010.

16. E. Scerri, Tales of technetium, *Nature Chemistry,* 1, 332, 2009. Also see my Recognizing rhenium, *Nature Chemistry,* 2, 589, 2010, and Finding francium, *Nature Chemistry,* 1, 670, 2009.

17. J. M. Keane, W. D. Harman, A New Generation of π –Basic Dearomatization Agents, *Organometallics,* 24, 1786–1798, 2005.

18. H. K. Yoshihara, Ogawa's Discovery of Nipponium and Its Re-evaluation, *Historia Scientiarum, 9,* 277–269, 2000; Ogawa Family and their Nipponium Research: H. K. Yoshihara, T. Kobayashi, M. Kaji, Successful Separation of the element 75 before Its Discovery by Noddacks, *Historia Scientarum, 15,* 177–190, 2005; H. K. Yoshihara, Nipponium as a new element (Z = 75) separated by the Japanese chemist, Masataka Ogawa: a scientific and science historical re-evaluation, *Proceedings of the Japanese Academy, Series B, 84,* 232–244, 2008.

19. M. Ogawa, *Chemical News,* 98, 249–251, 1908; M. Ogawa, *Chemical News, 98,* 261–264, 1908.

20. Yoshihara's translation: A few years later, Ogawa believed that he had found, in thorianite, the silicate of a new element, nipponium. Mr. R. B. Moore, the principal chemist of the Bureau of Mines in Washington, very obligingly sent us a few crystals of the silicate of nipponium obtained by Ogawa. These crystals are essentially composed of the zirconium silicate while having a 2-percent content of hafnium.

21. One must wonder how Yoshihara ascertained that his paper was "highly evaluated by many people." Did they all write messages of congratulations to him? Or is he referring to the one or two probable reviewers of the article? In

any case the mere reception of an article by undisclosed persons does little to add to the credibility of Yoshihara's claim.

22. The obvious question of why Ogawa could not personally have sent or delivered one of his samples to a foreign X-ray specialist is not contemplated by Yoshihara.

23. I learned of this fact in private correspondence with Yoshihara. I also find it odd that the way in which the calculated peaks were arrived at has not been explained in any of Yoshihara's published articles.

Chapter 6

1. I am grateful to Javier Garcia Martinez, an inorganic chemist at the University of Alicante in Spain, for pointing out to me that Mendeleev initially predicted four rather than three elements. Garcia Martinez was the designer of an attractive postage stamp issued by Spain to commemorate the hundredth anniversary of Mendeleev's death in 2007. Also see J. Garcia Martinez, P. Roman Polo, Spain Celebrates Its Year of Science, *Chemistry International*, 30, 4–8, 2008; D. Rabinovich, Mendeleev's Triumph, *Chemistry International*, vol. 29, July–August, 2007, p. 3.

2. As recently as the turn of the millennium, Belgian physicist Pieter van Assche collaborated with several US spectroscopists to reanalyze the original X-ray images. They then argued that the Noddacks had in fact isolated another element in addition to rhenium. But this claim has been disputed by a number of other researchers.

3. P. K. Kuroda, On the nuclear physical stability of the uranium minerals, *Journal of Chemical Physics*, 25, 781–782, 1956.

4. R. Bodu et al., Sur l'existence d'anomalies isotopiques rencontrés dan l'uranium du Gabon, *Comptes-rendus de l'Académie des sciences de Paris, D 275,* 1731–1736, 1972.

5. I am highly indebted to an article by William Griffith, Spurious Platinum Metals, *Chemistry in Britain, 4,* 430–435, 1968.

6. B. T. Kenna, The Search for Technetium in Nature, *Journal of Chemical Education, 39,* 436–442, 1962; R. Zingales, From Masurium to Trinacrium, *Journal of Chemical Education, 82,* 221–227, 2005. W. P. Griffith, Spurious Platinum Metals, *Chemistry in Britain, 4,* 430–444, 1868. The third of these authors considers polonium to be identical with ruthenium.

7. This is something that cannot happen these days. One of the naming rules framed by IUPAC is that after a name is first proposed and not accepted it can

not be used again. Unless this rule is ever revoked it means that no element will ever be named for Davy or Moseley, among others.

8. A good deal later, J. Newton Friend and Druce claimed that Kern might in fact have discovered rhenium some fifty years before the official date of the discovery or rather synthesis of this element. J. Newton Friend, J. G. F. Druce, Davyum a Possible Precursor of Rhenium, *Nature, 165,* 891–891, 1950.

9. Two other spurious metals, named josephinite and amarillium, were also linked to canadium. Needless to say these, too, were spurious.

10. M. Gerber, La recherche de deux metaux inconnus, *Le Moniteur Scientifique Quesneville,* Avril 1917.

11. A number of popular accounts on the discovery of rhenium imply that the Noddacks were the first to suggest that it might resemble elements that were horizontally rather than vertically adjacent. For example, Fathi Habashi's otherwise excellent account fails to note the priority of Gerber in this respect. Fathi Habashi, *Ida Noddack (1896–1978), Personal Recollections on the Occasion of 80th Anniversary of the Discovery of Rhenium,* Laval University Press, Laval, Canada, 2005.

12. Bosanquet, Keeley, *Philosophical Magazine,* (6), 145–147, 1924.

13. J. Newton Friend, *Man and the Chemistry of the Elements,* 2nd edition, Charles Griffin, London, 1961.

14. H. M. Van Assche, The Ignored Discovery of Element Z = 43, *Nuclear Physics A, A480,* 205–214, 1988.

15. Ibid.

16. Anonymous, The Disputed Discovery of Element 43. A Re-examination of an elegant early use of wavelength dispersive X-ray microanalysis, *Journal of Research of the National Institute of Standards and Technology, 104,* November–December, 599–599, 1999.

17. In addition, Van Assche and Armstrong gave a lecture at the National Meeting of the American Chemical Society in Anaheim, California, March 21–25, 1999. This lecture was attended by the present author, who was persuaded by the presentation. I also included an entry on technetium in my earlier book, *The Periodic Table, Its Story and Its Significance,* in which I erroneously accepted this work as evidence for the validity of the discovery by Noddack et al. in 1925.

18. It's Elemental, Special 80th Anniversary Issue of *Chemical & Engineering News,* September 8, 2003. Other articles that are relevant to the present book include one by the present author on hafnium (p. 138).

19. R. Zingales, From Masurium to Trinacrium: The Troubled Story of Element 43, *Journal of Chemical Education, 82,* 221–227, 2005.

20. F. Habashi, The History of Element 43—Technetium, *Journal of Chemical Education, 83,* 213–213, 2006; Habashi is also the author of a book dedicated to Ida Noddack. Fathi Habashi, *Ida Noddack (1896–1978),* Laval University Press, Laval, Quebec, 2005. P. Kuroda, A Note on the Discovery of Technetium, *Nuclear Physics A, 503,* 178–182, 1989.

21. B. T. Kenna, P. K. Kuroda, Isolation of naturally occurring technetium, *Journal of Inorganic and Nuclear Chemistry, 23,* 142–144, 1961.

22. R. Zingales, The History of Element 43—Technetium (the author replies), *Journal of Chemical Education, 83,* 213–213, 2006.

23. E. Segrè, Element 43, *Nature, 143,* 460–461, 1939.

24. What Paneth refers to here as a branch product is also referred to as actinium-K, or AcK, on his periodic table and in a later part of the article. It is the product obtained following α decay of actinium, $_{89}$Ac, as first fully characterized by Perey (see chapter 7).

25. F. Paneth, The Making of the Missing Chemical Elements, *Nature, 159,* 8–10, 1947.

26. The convener of the meeting was probably Paneth himself.

27. Although not nationalist in the usual sense, this comment is especially partisan given that Paneth is reminding the reader that Noddack was a sympathizer with the Nazi regime. Nor is it difficult to appreciate why Paneth may have felt a good deal of resentment. He was dismissed from his own position as professor of inorganic chemistry by the Nazis in 1936, after which he very soon fled to England, where he held positions at Imperial College and then Durham University. He did eventually return to Germany in 1953 to head the University of Mainz.

28. C. Perrier, E. Segrè, Technetium: the Element of Atomic Number 43, *Nature, 159,* 24–24, 1947; D. R. Corson, K. R. MacKenzie, E. Segrè, Astatine: the Element of Atomic Number 85, *Nature, 159,* 24–24, 1947.

29. E. Segrè, G. T. Seaborg, Nuclear isomerism in element 43, *Physical Review, 54,* 772–772, 1938.

30. Interview of E. Segrè by C. Weiner, B. Richman on February 13, 1937, Niels Bohr Library & Archives, American Institute of Physics, College Park, MD USA, http://www.aip.org/history/ohilist/4876.html; http://libserv.aip.org:81/ipac20/ipac.jsp?uri=full=3100001~!4875!0&profile=newcustom-icos

31. S. Flügge, Kann der Energieinhalt der Atomkerne technisch nutzbar gemacht werden? *Naturwissenschaften, 27,* 402–410, 1939.

32. G. A. Cowan, A Natural Fission Reactor, *Scientific American, 235*, 36–47, 1976.

33. In fact, naturally occurring technetium had previously been discovered by Kenna and the same Paul Kuroda, in 1961, prior to the discovery of the Oklo reactor. B. T. Kenna, P. K. Kuroda, Isolating Naturally Occurring Technetium, *Inorganic and Nuclear Chemistry, 23*, 142–144, 1961.

34. A. F. Holleman, E. Wiberg, *Inorganic Chemistry*, Academic Press, San Diego, CA, 2001.

Chapter 7

1. I. Langmuir, Pathological Science, *Physics Today*, 42, 36–48, 1989.

2. G. B. Kauffman, J. P. Adloff, Marguerite Perey and the Discovery of Francium, *Education in Chemistry*, September 1989, 135–137.

3. L. A. Orozco, Francium, *Chemical & Engineering News*, 2003, http://pubs.acs.org/cen/80th/francium.html

4. F. H. Loring, J. G. F. Druce, Eka-Caesium, *Chemical News, 31*, 289–289, 1925.

5. Ibid.

6. Ibid.

7. F. H. Loring, J. G. F. Druce, Eka-Caesium and Eka-Iodine, *Chemical News, 131*, 305–305, 1925.

8. Ibid.

9. Anonymous, Alabamine & Virginium, *Time Magazine*, Monday, Feb. 15, 1932.

10. The description of Allison's apparatus has been simplified in the present account.

11. H.G. MacPherson, The Magneto Optic Method of Chemical Analysis, *Physical Review, 47*, 310–315, 1934.

12. I could not discover whether Lewis got his $10.00 back.

13. This episode is recounted by Irving Langmuir in an interesting article that later appeared in *Physics Today* under the title of "Pathological Science" and that prominently featured Allison's method as a good example of what Langmuir also called "sick science." I. Langmuir, Pathological Science, *Physics Today*, 42, 36–48, 1989.

14. It was Perrin who provided evidence in favor of Einstein's statistical theory of Brownian motion, thus helping to establish the view that atoms "really exist."

15. H. Hulubei, The search for element 87, *Physical Review, 71,* 740–741, 1947.

16. J. P. Adloff, G. B. Kauffman, Francium (Atomic Number 87), the Last Discovered Natural Element, *The Chemical Educator,* 10, 2005.

17. Recall that β emission produces an element with one unit of atomic number higher.

18. M. Perey, Sur un element 87, dérivé de l'actinium, *Comptes-rendus hebdomadaires des séances de l'Académie des sciences, 208,* 97–98, 1939. Author's translation: According to the hypothesis that this radioelement is formed by α emission from actinium, it would occupy the place of element 87 in the periodic table; to prove this we have attempted to establish the chemical analogy of this body with ceasium by syncrystallization: the perchlorate of caesium having been chosen for this purpose because of the slight solubility that distinguishes it from the perchlorates of the nonalkaline metals, which are very soluble. On adding caesium chloride to the mother liquor and by precipitating with a solution of sodium perchlorate, some crystals formed that showed radioactivity, which decayed exponentially with a period of twenty-one minutes ± 1. Original quotation from page 98.

19. Ibid. Author's translation: We are therefore led to believe that this naturally radioactive element of 21 minute half-life, and atomic number 87, is formed by α decay of actinium given that actinium has a weak α branch or that it might be a mixture of two isotopes that disintegrates by β decay and the other by α decay. Original quotation from page 98.

20. A couple of years earlier Perrin had also presented an article by Hulubei entitled "Nouvelles recherches sur l'element 87 (Ml)" to the same academy.

21. In 1953, E. K. Hyde in the United States discovered that there is a weak α decay in ^{223}Fr, which produces element 85. This means that strictly speaking even element 85, or astatine, occurs naturally although its initial discovery was carried out via artificial synthesis.

22. The anapole moment has already been observed in Cs atoms. C. S. Wood, S. C. Bennett, D. Cho, B. P. Masterson, J. L. Roberts, C. E. Tanner, C. E. Wieman, Measurement of Parity Nonconservation and an Anapole Moment in Cesium, *Science,* 275, no. 5307, 1759–1763, 1997.

23. E. Gomez, L. A. Orozco, G. D. Sprouse, Spectrosocpy with trapped francium: perspectives for weak interaction studies, *Reports on Progress in Physics, 69,* 79–118, 2006.

24. L. A. Orozco, Francium, *Chemical & Engineering News, Special Issue on the Elements,* 2003; available at http://pubs.acs.org/cen/80th/francium.html

Chapter 8

1. B. F. Thornton, S. C. Burdette, Finding eka-iodine: Discovery Priority in Modern Times, *Bulletin for the History of Chemistry, 35,* 86–96, 2010.

2. H. G. MacPherson, The Magneto Optic Method of Chemical Analysis, *Physical Review, 47,* 310–315, 1935.

3. H. Hulubei, Mesures du spectre L du Ra (88), *Comptes Rendus des. Séances de l'Académie des Sciences. Serie C, 203,* 542–543, 1936; H. Hulubei, Emissions faibles dans le spectre L du Ra (88), *Comptes Rendus des Séances de l'Académie des Sciences, Serie C, 203,* 665–667, 1936.

4. M. Valadares, "Contributo Allo Studio Degli Spettri γ e X Molli dei Prodotti si Disintegrazione del Radon," *Rendiconti Instituto Sanita Publicà, 3,* 953–963, 1942; M. Valadares, "Gli Spettri γ e X dei derivati del radon nella regione UX 700 a 1300," *Rendiconti Instituto Sanita Publicà, 2,* 1049–1056, 1941.

5. B. Karlik and T. Bernert, Zur Frage eines Dualen Zerfalls des RaA, Sitzber. *Akad. Wiss. Wien, Math.-naturw. Klasse, 151,* 255–265, 1942; B. Karlik and T. Bernert, "Über eine Vermutete ß-Strahlung des Radium A und die Nat ürliche Existenz des Elementes 85, *Naturwissenschaften, 30,* 685–686, 1942.

6. According to Thornton and Burdette's article of 2010, Paneth's motivations may have been somewhat nationalistic. They claim that Paneth supported the Austrians because he had been a former colleague of theirs at the Institute for Radium Studies. Whereas he had been forced to leave Germany, the Austrians Karlik and Bernert had remained in the German territories, which of course included Austria, but Paneth was aware of Karlik's opposition to German war policies. In addition, they claim that Paneth was suspicious of the work of Hulubei and Cauchois, which was carried out in Nazi-occupied France, thus suggesting some form of compliance with German policies.

7. B. Karlik, Unsere Heutigen Kenntnisse über das Element 85 (Ekajod), *Monatshefte für Chemie, 77,* 348–351, 1947.

8. B. F. Thornton, S. C. Burdette, Finding eka-iodine: Discovery Priority in Modern Times, *Bulletin for the History of Chemistry, 35,* 86–96, 2010.

9. Ibid.

10. Private correspondence from the authors.

11. Element 85, Abstract, *Nature, 146,* 225–225, 1940.

12. Alice Leigh-Smith (née Prebil) was an English nuclear physicist. She married Philip Leigh-Smith, the son of the arctic explorer Benjamin Leigh-Smith, in 1933. She was a student of Marie Curie and the first British woman to receive a PhD in nuclear physics.

13. D. R. Corson, K. R. MacKenzie, E. Segrè, Artificially radioactive Element 85, *Physical Review*, *58*, 672–678, 1940.

14. K. Otozai, N. Takahashi, Estimation chemical form boiling point elementary astatine by radio gas chromatography, *Radiochimica Acta*, *31*, 201–203, 1982.

15. T. Jahn, MIPS and their role in the exchange of metalloids. *Advances in Experimental Biology and Medicine*. *679*, 41, 2010; A. G. Maddock, "Astatine." *Supplement to Mellor's comprehensive treatise on inorganic and theoretical chemistry, Supplement II, Part 1, (F, Cl, Br, I, At)*, 1064–1079, 1956; C. Housecroft, A. G. Sharpe, *Inorganic chemistry* (3rd ed.). Pearson Education, 2008, 533.

16. D. S. Wilbur, Overcoming the obstacles to clinical evaluation of ^{211}At-labeled radiopharmaceuticals, *The Journal of Nuclear Medicine*, *42*, 1516–1518, 2001; M. R. Zalutsky et al., High-level production of α-particle-emitting ^{211}At and preparation of ^{211}At-labeled antibodies for clinical use, *The Journal of Nuclear Medicine*, *42*, 1508–1515, 2001.

Chapter 9

1. The country later became Czechoslovakia and more recently, the Czech Republic.

2. Eventually, technetium was discovered in 1937.

3. In fact, Brauner claimed that dvi-manganese had recently been discovered by colleagues in his own institute. This is what he wrote in the letter to *Nature*: "As regards element No. 61, the difference between the atomic weights of Sm–Nd = 6.1, is greater than that between any other neighbouring elements. It is remarkable that it is of the same order as that between the atomic weights of Mo–Ru = 5.7, between which stands ekamanganese, and of Os–W = 6.9, between which stands dwi-manganese, recently discovered in our laboratory by Heyrovsky and Dolejsek."

4. André-Louis Debierne, a French chemist, announced the discovery of a new element in 1899. He separated it from pitchblende residues left by Marie and Pierre Curie after they had extracted radium. Debierne described the substance (in 1899) as similar to titanium and (in 1900) as similar to thorium. Friedrich Oskar Giesel independently discovered actinium in 1902 as a substance that was similar to lanthanum and called it "emanium" in 1904. After

a comparison of substances in 1904, Debièrne's name was retained because it had seniority. The history of the discovery of actinium remained questionable for decades. Articles published in the 1970s and later suggest that Debièrne's results published in 1904 conflict with those reported in 1899 and 1900. Whether Debièrne and Giesel should share the merit of discovery or if Giesel alone should be credited with the discovery is still under debate.

5. The five pair reversals consist of K and Ar, Co and Ni, Te and I, and Th and Pa, U and Np. Eric Scerri, *A Very Short Introduction to the Periodic Table,* Oxford University Press, Oxford, 2011.

6. Author's translation: The material available to us was such a small quantity that we did not think it scrupulous to publish our findings and so we sent a small package containing our results, and the photos of the spectra that are under discussion, to the Academy of the Lincei.

7. M. Costa, M. Fotani, P. Manzelli, P. Papini, Storia della scoperta dell'elemento 61, in Storia e Fondamenti della Chimica, *Memorie di Scienze Fisiche e Naturali,* 1997, 441–442.

8. Author's translation: We hypothesize that the samples analyzed in Florence and America could have contained minimal traces of this element. One of the authors, Marco Fontani, has written to me and verified that this is a correct translation. He also writes that the phrase "minimal traces" was meant to refer to amounts that were undetectable either in the 1920s or as recently as the 1990s when this article was written. It should also be emphasized how the contemporary Italian authors show a refreshingly nonnationalistic attitude in crediting the American chemists to the same extent as their Italian compatriots.

9. My account is much indebted to a recent article by Clarence B. Murphy, who has had a long-standing interest in the discovery of element 61 as well as access to many archival sources belonging to both James and Smith Hopkins. C. J. Murphy, Charles James, B. Smith Hopkins, and the tangled web of element 61, *Bulletin for the History of Chemistry, 31,* 9–18, 2006.

10. This raises questions about Brauner's originality in recognizing this gap between atomic weights.

11. W. F. Peed, K. J. Spitzer, and L. E. Burkhart, The L Spectrum of Element 61, *Phyical Review 76,* 143–144, 1949.

12. In his extended historical paper, which is very favorable to the claims of James et al., Clarence Murphy also fails to take a position on whether or not James had detected element 61, although he does say, "It is striking that the six lines reported by James and the two by Hopkins are remarkably close to those determined from an authentic sample of the element 61."

ıs, and the tangled web of
9–18, 2006.

lements were published in

pril 26, 1925, and published
was published in December

specialist working at the
ble source, who spent many
following joke was current
spected of having messed up
orked it up!"

omic bomb, one of the most
aration of certain rare earth
ıportant isotopes of uranium

, The Chemical Identification
nt 61, *Journal of the American*

m a Lanthanon Mixture, *Acta*
, 1965.

tchblende, *Journal of Inorganic*

21. A. Kavetskiy, G. Yakubova, S. M. Yousaf, K. Bower, J. D. Robertson, and A. Garnov, Efficiency of Pm-147 Direct Charge Radioisotope Battery, *Applied Radiation and Isotopes, 69*, 744–748, 2011.

22. One of the leaders in the development of ^{147}Pm batteries is J. David Robertson at the Chemistry Department at the University of Missouri. In an interview he told me that his laboratory currently holds the largest amount of promethium anywhere in the United States.

Chapter 10

1. When Mendeleev published his periodic tables, there were of course many gaps. This book has focused on filling the last remaining 7 gaps in the old periodic table consisting of elements 1–92. Even before the last of these gaps had been plugged, it was realized that the actinide elements were part of a 32 element series, of which most members were yet to be synthesized. Only in

2010 was the seventh period of the table, consisting of 32 elements, finally completed.

2. Ernest Rutherford, Collisions of alpha Particles with Light Atoms. IV. An Anomalous Effect in Nitrogen, *The London, Edinburgh and Dublin Philosophical Magazine and Journal of Science*, 6th series, 37, 581–586 (1919).

3. Heilbron, J. L.; Robert W. Seidel (1989). *Lawrence and His Laboratory: A History of the Lawrence Berkeley Laboratory*. Berkeley: University of California Press.

4. Chadwick, J., Possible Existence of a Neutron, *Nature, 129*, 312, 1932.

5. E. Fermi, Possible Production of Elements of Atomic Number Higher than 92, *Nature, 133*, 898–899, 1934. Enrico Fermi's (1901–1954) colleagues were Edoardo Amaldi (1908–1989), Oscar D'Agostino (1901–1975), Emilio Segrè (1905–1989), and Franco Rasetti (1901–2001). The Dean of the Faculty of Rome University, Orso Mario Corbino (1876–1937), announced the discovery of the elements 93 and 94 and prematurely gave them the names and symbols Ausonium, Ao, after Ausonia, the poetic name of Italy, and Hesperium (Esperio), from Hesperius, the Western country (Italy, seen from Greece). Meanwhile, the then fascist regime of Italy wanted him to call one of these elements Littorio (Littorium, after the Italian "littorio," an Imperial Roman symbol used during the Mussolini dictatorship). Corbino sarcastically replied that it was unlucky for the regime to be associated with an element with a half-life of a few seconds, whereupon the names remained Ausonium and Hesperium. My thanks to Marco Fotani for this information.

6. E. McMillan, P. Abelson, Radioactive Element 93, *Physical Review 57*, 12, 1185, 1950.

7. P. R. Fields et al., Transplutonium Elements in Thermonuclear Test Debris. *Physical Review 102*, 180–182, 1956.

8. A. Ghiorso, M. Nurmia, J. Harris, K. Eskola, P. Eskola, Positive Identification of Two Alpha-Particle-Emitting Isotopes of Element 104, *Physical Review Letters, 22*, 1317–1320, 1969.

9. M. Fleischmann, S. Pons, Electrochemically induced nuclear fusion of deuterium, *Journal of Electroanalytical Chemistry 261*, (2A), 301–308, 1989.

10. G. Münzenberg, G. S. Hofmann et al., Identification of element 107 by α correlation chains, *Zeitschrift für Physik A 300*: 107, 1981; G. Münzenberg et al., The identification of element 108, *Zeitschrift für Physik A 317*, 235, 1984; G. Münzenberg, P. Armbruster et al., Observation of one correlated α-decay in the reaction ^{58}Fe on ^{209}Bi$\rightarrow^{267}$109, *Zeitschrift für Physik A 309*, 89, 1982.

11. S. Hofmann, V. Ninov et al., Production and decay of 269110, *Zeitschrift für Physik A 350*, 277, 1995.

12. S. Hofmann, V. Ninov et al., The new element 111. *Zeitschrift für Physik A 350*, 281, 1995.

13. S. Hofmann et al., The new element 112, *Zeitschrift für Physik A 354*, 229–230, 1996; also see, P. Armbruster, F. P. Hessberger, Making New Elements, *Scientific American*, 279, 72–77, Sept. 1998.

14. Y. Oganessian et al. Synthesis of superheavy nuclei in the ^{48}Ca+^{244}Pu reaction: *Physical Review C 62* (4): 041604, 2000; Y. Oganessian et al., Experiments on the synthesis of element 115 in the reaction. *Physical Review C 69* (2): 021601, 2004; Y. Oganessian et al. Observation of the decay of ^{292}116. *Physical Review C 63*: 011301, 2000.

15. A. V. Yeremin et al., Synthesis of nuclei of the superheavy element 114 in reactions induced by ^{48}Ca, *Nature 400*, 242–245, 1999.

16. N. D. Cooke, *Models of the Atomic Nucleus*, Springer, Berlin, 2010.

17. V. Ninov et al., Observation of Superheavy Nuclei Produced in the Reaction of 86Kr with 208Pb. *Physical Review Letters 83* (6): 1104–1107, 1999.

18. S. LeVay, *When Science Goes Wrong*, Penguin, New York, 2008, chapter 12.

19. Y. Oganessian et al., Synthesis of the Isotopes of Elements 118 and 116 in the Cf-249 and Cm-245+Ca-48 Fusion Reactions, *Physical Review C* 74 (4), 044602, 2006.

20. P. Pyykkö, A suggested periodic table up to Z ≤ 172, based on Dirac–Fock calculations on atoms and ions. *Phys. Chem. Chem. Phys.*, *13*, 161–168, 2011.

21. Ibid.

22. A. Türler et al., Evidence for Relativistic Effects in the Chemistry of Element 104, *The Journal of Alloys and Compounds*, 271–273, 287–291, 1998; D. Hoffman, The Heaviest Elements, *Chemical & Engineering News*, May 2, 1994, 24–34.

23. R. Lougheed, Oddly ordinary seaborgium, *Nature*, *338*, 21–21, 1997; R. Eichler et al., Chemical Characterization of bohrium (element 107), *Nature*, *407*, 63–65, 2000.

24. See for example my own, E. R. Scerri, *A Very Short Introduction to the Periodic Table*, Oxford University Press, 2011, chapter 9.

25. R. Eichler et al., Chemical Characterization of Bohrium (element 107), *Nature*, *407*, 63–65, 2000.

26. R. Eichler et al., Chemical Characterization of element 112, *Nature, 447,* 72–75, 2007.

27. P. Pyykkö, Relativistic Effects in Structural Chemistry, *Chemical Reviews, 88.* 563–594, 1988; Relativity, Gold, Closed-Shell Interactions, and CsAu. NH3, *Angewandte Chemie International Edition, 41,* 3573–3578, 2002.

28. P. Schwerdtfeger, M. Seth, Relativistic Effects of the Superheavy Elements, in *The Encyclopedia of Computational Chemistry,* eds P. v. R. Schleyer, N. L. Allinger, T. Clark, J. Gasteiger, H. F. Schaefer III, P. R. Schreiner, John Wiley and Sons, New York, 1998.

BIBLIOGRAPHY

J. P. Adloff, G. B. Kauffman, Francium (Atomic Number 87), the Last Discovered Natural Element, *The Chemical Educator, 10,* 2005.

H. Aldersey-Williams, *Periodic Tales,* Penguin Viking, London, 2011.

Anonymous, The Disputed Discovery of Element 43. A Re-examination of an elegant early use of wavelength dispersive X-ray microanalysis, *Journal of Research of the National Institute of Standards and Technology, 104,* November–December, 599–599, 1999.

P. Armbruster, F. P. Hessberger, Making New Elements, *Scientific American.* Sept. 1998, 72–77.

M. Arrtep, P. K. Kuroda, Promethium in Pitchblende, *Journal of Inorganic Nuclear Chemistry, 30,* 699, 1968.

R. M. Baum, ed., It's Elemental, Special 80th Anniversary Issue of *Chemical & Engineering News,* September 8, 2003.

E. Béguyer De Chancourtois, Vis Tellurique, *Compes Rendus de l'Académie des Sciences, 54,* 1862, 757–761, 840–843, 967–971.

R. Bodu et al., Sur l'existence d'anomalies isotopiques rencontrés dan l'uranium du Gabon, *Comptes-rendus de l'Académie des sciences de Paris, D 275,* 1731–1736, 1972.

N. Bohr, On the Constitution of Atoms and Molecules, *Philosophical Magazine, 26,* 1–25, 476–502, 857–875, 1913 (known as the trilogy paper).

C. H. Bosanquet, T. C. Keeley, Note on the search for element 43, *Philosophical Magazine, 48,* 145–147, 1924.

A. Brannigan, *The Social Basis of Scientific Discoveries,* Cambridge University Press, Cambridge, 1981.

W. H. Brock, *William Crookes (1832–1919) and the Commercialization of Science*, Ashgate, Aldershot, UK, 2008.

J. Chadwick, Possible Existence of a Neutron, *Nature, 129*, 312, 1932.

N. D. Cooke, *Models of the Atomic Nucleus*, Springer, Berlin, 2010.

D. R. Corson, K. R. MacKenzie, E. Segrè, Artificially radioactive Element 85, *Physical Review, 58*, 672–678, 1940.

D. R. Corson, K. R. MacKenzie, E. Segrè, Astatine: The Element of Atomic Number 85, *Nature, 159*, 24–24, 1947.

M. Costa, M. Fotani, P. Manzelli, P. Papini, Storia della scoperta dell'elemento 61, in Storia e Fondamenti dellaChimica, *Memorie di Scienze Fisiche e Naturali*, 1997, 441–442.

D. Coster and G. Hevesy, On the New Element Hafnium, *Nature, 111*, 462–463, 1923.

F. A. Cotton, R. A. Walton, *Multiple Bonds Between Metal Atoms*, Oxford University Press, Oxford, 1993.

G. A. Cowan, A Natural Fission Reactor, *Scientific American*, 235, 36–47, *1976*.

W. Crookes, Radio-Activity of Uranium, *Proceedings of the Royal Society of London, 66*, 409–423, 1899–1900.

J. E. Earley, How chemistry shifts horizons: element, substance, and the essential, *Foundations of Chemistry, 11*, 65–77, 2009.

R. Eichler et al., Chemical Characterization ofbohrium (element 107), *Nature, 407*, 63–65, 2000.

R. Eichler et al., Chemical Characterization of element 112, *Nature, 447*, 72–75, 2007.

A. Einstein, *Investigations of the Theory of the Brownian Movement*, with notes by R. Fürth, translated by A. D. Cowper, Dover Publications, Mineola, New York, 1956.

J. Emsley, *The Elements*, 3rd Edition, Clarendon Press, Oxford, 1998.

J. Emsley, *The A–Z of the Elements*, Oxford University Press, Oxford, 2001.

O. Erämetsä, Separation of Prometheum from a Lanthanon Mixture, *Acta Polytechnica. Scandinavica Chem. Mat. Sci, 37*, 21, 1965.

P. R. Fields et al., Transplutonium Elements in Thermonuclear Test Debris, *Physical Review 102*, 180–182, 1956.

M. Fleischmann, S. Pons, Electrochemically induced nuclear fusion of deuterium, *Journal of Electroanalytical Chemistry 261*, (2A), 301–308, 1989.

S. Flügge, Kann der Energieinhalt der Atomkerne technisch nutzbar gemacht werden? *Naturwissenschaften, 27*, 402–410, 1939.

J. Garcia Martinez, P. Roman Polo, Spain Celebrates Its Year of Science, *Chemistry International, 30*, 4–8, 2008.

M. Gerber, La recherche de deux metaux inconnus, *Le Moniteur Scientifique de Quesneville*, Avril 1917.

A. Ghiorso, M. Nurmia, J. Harris, K. Eskola, P. Eskola, Positive Identification of Two Alpha-Particle-Emitting Isotopes of Element 104, *Physical Review Letters, 22*, 1317–1320, 1969.

E. Gomez, L. A. Orozco, G. D. Sprouse, Spectroscopy with trapped francium: perspectives for weak interaction studies, *Reports on Progress in Physics, 69*, 79–118, 2006.

W. P. Griffith, Spurious Platinum Metals, *Chemistry in Britain, 4*, 430–444, 1868.

A. G. Gross, Do Disputes over Priority Tell Us Anything about Science? *Science in Context, 11*, 161–179, 1998.

A. V. Grosse, The Analytical Chemistry of Element 91, *Journal of the American Chemical Society, 52*, 1742–1747, 1930.

F. Habashi, *Ida Noddack (1896–1978), Personal Recollections on the Occasion of 80th Anniversary of the Discovery of Rhenium*, Laval University Press, Laval, Canada, 2005.

F. Habashi, The History of Element 43—Technetium, *Journal of Chemical Education, 83*, 213–213, 2006.

O. Hahn, L. Meitner, Die Muttersubstanz des Actiniums, ein Neues Radioaktives Element von Langer Lebensdauer. *Physikalische Zeitschrift, 19*, 208–218, 1918.

O. Hahn and F. Strassmann, *Die Naturwissenschaften 27*, p. 11–15 (January 1939), received December 22, 1938.

D. C. Hamilton, M. A. Jensen, Mechanism for Superconductivity in lanthanum and uranium, *Physical Review Letters, 11*, 205, 1963.

D. C. Hamilton, Position of lanthanum in the periodic table, *American Journal of Physics*, 33, 637, 1965.

H. M. Hansen, S. Werner, Optical Spectrum of Hafnium, *Nature, 111*, 322–322, 1923.

P. J. Hartog, A First Foreshadowing of the Periodic Law, *Nature, 41*, 186–188, 1889.

R. F. Hendry, *Lavoisier and Mendeleev on the Elements, Foundations of Chemistry*, 7, 31–48, 2005.

J. L. Heilbron, R. W. Seidel, *Lawrence and His Laboratory: A History of the Lawrence Berkeley Laboratory*. Berkeley: University of California Press, 1989.

G. Hevesy, *Adventures in Radioisotope Research*, Pergamon Press, Oxford, 1962.

G. Hinrichs, *The Elements of Chemistry and Mineralogy*, Griggs, Watson & Day, Davenport, Iowa, 1871.

G. Hinrichs, *The Principles of Chemistry and Molecular Mechanics*, Day, Egbert & Fidlar, Davenport, Iowa, 1874.

D. Hoffman, The Heaviest Elements, *Chemical & Engineering News*, May 2, 1994, 24–34.

R. Hoffmann, C. Djerassi, *Oxygen, a play*, Wiley-VCH, 2001.

S. Hofmann, V. Ninov et al., Production and decay of ^{269}110, *Zeitschrift für Physik A 350*, 277, 1995.

S. Hofmann, V. Ninov et al., The new element 111, *Zeitschrift für Physik A 350*, 281, 1995.

S. Hofmann et al., The new element 112, *Zeitschrift für Physik A354*, 229–230, 1996; also see, P. Armbruster, F. P. Hessberger, Making New Elements, *Scientific American, 279*, 72–77, Sept. 1998.

A. F. Holleman, E. Wiberg, *Inorganic Chemistry*, Academic Press, San Diego, CA, 2001.

C. Housecroft, A. G. Sharpe, *Inorganic chemistry* (3rd ed.). Pearson Education, Prentice-Hall, New Jersey, 2008.

H. Hulubei, Mesures du spectre L du Ra (88), *Comptes Rendus des Séances de l'Académie des Sciences Serie C, 203*, 542–543, 1936.

H. Hulubei, Emissions faiblesdans le spectre L du Ra (88), *Comptes Rendus des Séances de l'Académie des Sciences, Serie C, 203*, 665–667, 1936.

H. Hulubei, The search for element 87, *Physical Review, 71*, 740–741, 1947.

T. Jahn, MIPS and their role in the exchange of metalloids, *Advances in Experimental Biology and Medicine. 679*, 41–203, 2010.

W. B. Jensen, Classification, Symmetry and the Periodic Table, *Computation and Mathematics with Applications, 12B*, 487–509, 1986.

B. Karlik and T. Bernert, Zur Frage eines Dualen Zerfalls des RaA, Sitzber, *Akad. Wiss. Wien, Math.-naturw. Klasse, 151*, 255–265, 1942.

B. Karlik and T. Bernert, Über eine Vermuteteß-Strahlung des Radium A und die Natürliche Existenz des Elementes 85, *Naturwissenschaften, 30*, 685–686, 1942.

B. Karlik, Unsere Heutigen Kenntnisse über das Element 85 (Ekajod), *Monatsheft e für Chemie, 77*, 348–351, 1947.

V. Karpenko, The Discovery of Supposed New Elements, *Ambix, 27*, 77–102, 1980.

G. B. Kauffman, J. P. Adloff, Marguerite Perey and the Discovery of Francium, *Education in Chemistry, 26*, 135–137, 1989.

A. Kavetskiy, G. Yakubova, S. M. Yousaf, K. Bower, J. D. Robertson, and A. Garnov, Efficiency of Pm-147 Direct Charge Radioisotope Battery, *Applied Radiation and Isotopes, 69*, 744–748, 2011.

S. Kean, *The Disappearing Spoon*, Little, Brown and Co., New York, 2010.

J. M. Keane, W. D. Harman, A New Generation of π- Basic Dearomatization Agents, *Organometallics, 24,* 1786–1798, 2005.

B. T. Kenna, P. K. Kuroda, Isolating Naturally Occurring Technetium, *Inorganic and Nuclear Chemistry, 23,* 142–144, 1961.

B. T. Kenna, The Search for Technetium in Nature, *Journal of Chemical Education, 39,* 436–442, 1962.

M. A. Korzhinsky, S. I. Tkachenko, K. I. Shmulovich, Y. A. Taran, G.S. Steinberg, Discovery of a pure rhenium mineral at Kudriavy volcano, *G. S. Nature 369*: 51–52, 2004.

H. Kragh, Elements No. 70, 71, and 72: Discoveries and Controversies, in C. H. Evans (Ed.), *Episodes in the History of Rare Earth Elements,* 67–89, Kluwer Academic Publishers, 1996.

H. Kragh, P. Robertson, On the Discovery of Element 72, *Journal of Chemical Education, 56,* 456–459, 1979.

H. Kragh, Conceptual Changes in Chemistry: The Notion of a Chemical Element, ca. 1900–1925, *Studies in History and Philosophy of Modern Physics, 31,* 435–450, 2000.

T. S. Kuhn, Historical Structure of Scientific Discovery, *Science, 136,* 760–764, 1962.

T. S. Kuhn, *Historical Structure of Scientific Revolutions,* 2nd ed., University of Chicago Press, Chicago, 1970, p. 55.

P. K. Kuroda, On the nuclear physical stability of the uranium minerals, *Journal of Chemical Physics, 25,* 781–782, 1956.

P. K. Kuroda, A Note on the Discovery of Technetium, *Nuclear Physics, A503,* 178–182, 1989.

I. Langmuir, Pathological Science, *Physics Today, 42,* 36–48, 1989.

S. J. La Placa, B. Post, The Crystal Structure of rhenium diboride, *Acta Crystallographica, 15,* 97, 1962.

P. E. Lecoq De Boisbaudran, A. Lapparent, A Reclamation of Priority on Behalf of M. De Chancourtois Referring to the Numerical Relations Among Atomic Weights, *Chemical News, 63,* 51–52, 1891.

S. LeVay, *When Science Goes Wrong,* Penguin, New York, 2008.

J. Levy, *Scientific Feuds,* New Holland, London, 2010.

F. H. Loring, J. G. F. Druce, Eka-Caesium, *Chemical News, 131,* 289–289, 1925.

F. H. Loring, J. G. F. Druce, Eka-Caesium and Eka-Iodine, *Chemical News, 131,* 305–305, 1925.

J. Lothar Meyer in Oswald's *Klassiker der Exacten Wissenschaften*: 30, Arbis eines Lehrganges der theoretischen Chemie, vorgetragen von Prof. S. Cannizzaro, Leipzig, 1891.

R. Lougheed, Oddly ordinary seaborgium, *Nature, 338*, 21–21, 1997.

S. Lyle, Narrative understanding: developing a theoretical context for understanding how children make meaning in classroom settings, *Journal of Curriculum Studies, 32*, 45–63, 2000.

H. G. MacPherson, The Magneto-Optic Method of Chemical Analysis, *Physical Review, 47*, 310–315, 1935.

A. G. Maddock, Astatine, *Supplement to Mellor's comprehensive treatise on inorganic and theoretical chemistry, Supplement II, Part 1, (F, Cl, Br, I, At)*, 1064–1079, 1956.

A. Marinsky, L. E. Glendenin, C. D. Coryell, The Chemical Identification of Radioisotopes of Neodymium and of Element, *Journal of the American Chemical Society, 69*, 2781–2785, 1947.

G. Markus, Why Is There No Hermeneutics of Natural Sciences? Some Preliminary Theses, *Science in Context, 1*, 5–15, 1987.

S. McGrayne, *Nobel Prize Women in Science*, John Henry Press, Washington, D.C., 2002.

E. McMillan, P. Abelson, Radioactive Element 93, *Physical Review, 57*, 12, 1185, 1950.

R. Merton, Priorities in Scientific Discovery, *American Sociological Review, 22*, 635–659, 1957.

H. Merz, K. Ulmer, Positions of lanthanum and lutetium in the Periodic Table, *Physics Letters, 26A*, 6–7, 1967.

H. G. J. Moseley, Atomic Models and X-Ray Spectra, *Nature, 92*, 544–544, 1913.

M. Mulkay, Norms and Ideology in Science, *Social Science Information, 15*, 637–656, 1976.

G. Münzenberg, S. Hofmann et al., Identification of element 107 by α correlation chains, *Zeitschrift für Physik, A300*, 107, 1981.

G. Münzenberg et al., The identification of element 108, *Zeitschrift für Physik, A317*, 235, 1984.

G. Münzenberg, P. Armbruster et al., Observation of one correlated α-decay in the reaction ^{58}Fe on $^{209}Bi \rightarrow {}^{267}109$, *Zeitschrift für Physik, A309*, 89–90, 1982.

C. J. Murphy, Charles James, B. Smith Hopkins, and the tangled web of element 61, *Bulletin for the History of Chemistry, 31*, 9–18, 2006.

P. Needham, Has Daltonian Atomism Provided Chemistry with Any Explanations? *Philosophy of Science, 71*, 1038–1047, 2004.

P. Needham, When Did Atoms Begin To Do Any Explanatory Work in Chemistry? *International Studies in the Philosophy of Science, 18*, 199–219, 2004.

J. A. R. Newlands, On the Law of Octaves, *Chemical News, 12*, 83–83, August 18, 1865.

J. A. R. Newlands, On the Law of Octaves, *Chemical News, 13*, 130–130, 1866.

J. Newton Friend, J. G. F. Druce, Davyum a Possible Precursor of Rhenium, *Nature, 165*, 891, 1950.

J. Newton Friend, *Man and the Chemistry of the Elements*, 2nd edition, Charles Griffin, London, 1961.

V. Ninov et al., Observation of Superheavy Nuclei Produced in the Reaction of ^{86}Kr with ^{208}Pb, *Physical Review Letters 83* (6), 1104–1107, 1999.

W. Noddack, I. Tacke, O. Berg, Die Ekamangane, *Naturwissenschaften 13* (26), 567–574, 1925.

W. Odling, On the Proportional Numbers of the Elements, *Quarterly Journal of Science, 1*, 642–648, October 1864.

Y. Oganessian et al., Synthesis of superheavy nuclei in the ^{48}Ca+^{244}Pu reaction, *Physical Review C 62* (4), 041604, 2000.

Y. Oganessian et al., Observation of the decay of ^{292}116. *Physical Review C 63*, 011301, 2000.

Y. Oganessian et al., Experiments on the synthesis of element 115 in the reaction. *Physical Review C 69* (2), 021601, 2004.

Y. Oganessian et al., Synthesis of the Isotopes of Elements 118 and 116 in the Cf-249 and Cm-245+Ca-48 Fusion Reactions, *Physical Review C 74* (4), 2006.

M. Ogawa, *Chemical News 98*, 249–251, 1908.

M. Ogawa, *Chemical News 98*, 261–264, 1908.

L. A. Orozco, Francium, *Chemical & Engineering News, Special Issue on the Elements*, 2003; available at http://pubs.acs.org/cen/80th/francium.html

K. Otozai, N. Takahashi, Estimation chemical form boiling point elementary astatine by radio gas chromatography, *Radiochimica Acta, 31*, 201–203, 1982.

F. Paneth, The Making of the Missing Chemical Elements, *Nature, 159*, 8–10, 1947.

W. F. Peed, K. J. Spitzer, and L. E. Burkhart, The L Spectrum of Element 61, *Physical Review 76*, 143–144, 1949.

M. Perey, Sur un element 87, dérivé de l'actinium, *Comptes-rendus hebdomadaires des séances de l'Académie des sciences, 208*, 97–98, 1939.

C. Perrier, E. Segrè, Technetium: the Element of Atomic Number 43, *Nature, 159*, 24–24, 1947.

J. Perrin, Mouvement brownien et réalité moléculaire, *Annales de Chimie et de Physique, 18*, 1–114, 1909.

K. R. Popper, *The Open Universe,* London, Hutchinson, ed. W. W. Bartley III., 1982, pp. 163–164.

P. Pyykkö, Relativistic Effects in Structural Chemistry, *Chemical Review, 88,* 563–594, 1988; Relativity, Gold, Closed-Shell Interactions, and $CsAu.NH_3$, *Angewandte Chemie International Edition, 41,* 3573–3578, 2002.

P. Pyykkö, A suggested periodic table up to Z ≤ 172, based on Dirac–Fock calculations on atoms and ions. *Physical Chemistry Chemical Physics, 13,* 161–168, 2011.

J. Qin et al., Is Rhenium Diboride a Superhard Material? *Advances in Materials, 20,* 4780, 2008.

D. Rabinovich, Mendeleev's Triumph, *Chemistry International,* vol. 29, 3, July–August 2007.

E. Rancke-Madsen, The Discovery of an Element, *Centaurus, 19,* 299–313, 1976.

G. Rayner-Canhan, Isodiagonality in the Periodc Table, *Foundations of Chemistry, 13,* 121–129, 2011.

T. Rothman, *Everything's Relative,* Wiley, Hoboken, NJ, 2003.

E. Rutherford, Collisions of alpha Particles with Light Atoms. IV. An Anomalous Effect in Nitrogen, *The London, Edinburgh and Dublin Philosophical Magazine and Journal of Science,* 6th series, *37,* 581–586 (1919).

E. R. Scerri, The Evolution of the Periodic System, *Scientific American, 279,* 78–83, 1998.

E. R. Scerri, *The Periodic Table, Its Story and Its Significance,* Oxford University Press, New York, 2007.

E. R. Scerri, Collected Papers in Philosophy of Chemistry, Imperial College Press, London, 2008.

E. R. Scerri, Which Element Belongs in Group 3?, *Journal of Chemical Education, 86,* 1188–1188, 2009.

E. R. Scerri, Tales of technetium, *Nature Chemistry, 1,* 332, 2009. Also see my Recognizing rhenium, *Nature Chemistry, 2,* 589, 2010.

E. R. Scerri, Finding francium, *Nature Chemistry, 1,* 670, 2009.

E. R. Scerri, *A Very Short Introduction to the Periodic Table,* Oxford University Press, Oxford, 2011.

E. R. Scerri, What is an element? What is the periodic table? And what does quantum mechanics contribute to the question? *Foundations of Chemistry, 14,* 69–81, 2012.

W. H. E. Schwarz, The Full Story of the Electron Configurations of the Transition Elements, *Journal of Chemical Education, 87,* 444–448, 2010.

P. Schwerdtfeger, M. Seth, Relativistic Effects of the Superheavy Elements, in *The Encyclopedia of Computational Chemistry* eds P. v. R. Schleyer, N. L. Allinger, T. Clark, J. Gasteiger, H. F. Schaefer III, P. R. Schreiner, John Wiley and Sons, New York, 1998.

E. Segrè, G. T. Seaborg, Nuclear isomerism in element 43, *Physical Review, 54*, 772–772, 1938.

E. Segrè, Element 43, *Nature, 143*, 460–461, 1939.

R. Sime, *Lise Meitner, A Life in Physics*, University of California Press, Berkeley, CA, 1996.

J. R. Smith, *Persistence and Periodicity*, unpublished PhD thesis, University of London, 1975.

A. Stwertka, *Guide to the Elements*, Oxford University Press, New York, 1998.

B. F. Thornton, S. C. Burdette, Finding eka-iodine: Discovery Priority in Modern Times, *Bulletin for the History of Chemistry, 35*, 86–96, 2010.

P. Thyssen, *Accommodating the Rare-Earths in the Periodic Table*, M.Sc. Thesis, Catholic University of Leuven, 2009.

A. Türler et al., Evidence for Relativistic Effects in the Chemistry of Element 104, *The Journal of Alloys and Compounds*, 271–273, 287–291, 1998.

G. Urbain, Sur le celtium, élément de numéroatomique 72, *Comptes Rendus de l'Académie des Sciences, 176*, 469–470, 1923.

M. Valadares, Contributo Allo Studio Degli Spettri γ e X Molli dei Prodotti di Disintegrazione del Radon, *Rendiconti Istituto Sanita Publicà, 3*, 953–963, 1940.

M. Valadares, Gli spettri γ e X deiderivati del radon nella regione UX 700 a 1300, *Rendiconti R. Accademia D'Italia, 2*, 1049–1056, 1941.

H. M. Van Assche, The Ignored Discovery of Element Z = 43, *Nuclear Physics A480*, 205–214, 1988.

A. J. van den Broek, The α Particle and the Periodic System of the Elements, *Annalen der Physik, 23*, 199–203, 1907.

A. J. van den Broek, The Number of Possible Elements and Mendeléeff's "Cubic" Periodic System, *Nature, 87*, 78–78, 1911.

S. Weinberg, Reductionism Redux, *The New York Review of Books*, October 5, 1995. Reprinted in S. Weinberg, *Facing Up*, Harvard University Press, 2001.

C. Weiner, B. Richman, Interview of Segrè, on February 13, 1937, Niels Bohr Library & Archives, American Institute of Physics, College Park, MD.

D. S. Wilbur, Overcoming the obstacles to clinical evaluation of ^{211}At-labeled radiopharmaceuticals, *The Journal of Nuclear Medicine, 42*, 1516–1518, 2001.

R. E. Wilson, Peculiar Protactinium, *Nature Chemistry*, *4*, 586–586, 2012.

C. S. Wood, S. C. Bennett, D. Cho, B. P. Masterson, J. L. Roberts, C. E. Tanner, C. E. Wieman, Measurement of Parity Nonconservation and an Anapole Moment in Cesium, *Science*, *275* no. 5307, 1759–1763, 1997.

A. V. Yeremin et al., Synthesis of nuclei of the superheavy element 114 in reactions induced by ^{48}Ca, *Nature 400*, 242–245, 1999.

H. K. Yoshihara, Ogawa's Discovery of Nipponium and Its Re-evaluation, *Historia Scientiarum*, *9*, 257–269, 2000.

H. K. Yoshihara, Ogawa Family and Their Nipponium Research: H. K. Yoshihara, T. Kobayashi, M. Kaji, Successful Separation of the element 75 before Its Discovery by Noddacks, *Historia Scientiarum*, *15*, 177–190, 2005.

H. K. Yoshihara, Nipponium as a new element (Z = 75) separated by the Japanese chemist, Masataka Ogawa: a scientific and science historical re-evaluation, *Proceedings of the Japanese Academy, Series B*, *84*, 232–244, 2008.

H. K. Yoshihara, Discovery of a new element 'nipponium': re-evaluation of pioneering works of Masataka Ogawa and his son Eijiro Ogawa, *Spectrochimica Acta Part B Atomic Spectroscopy*, *59*, 1305–1310, 2004.

M. R. Zalutsky et al., High-level production of -particle-emitting ^{211}At and preparation of ^{211}At-labeled antibodies for clinical use, *The Journal of Nuclear Medicine*, *42*, 1508–1515, 2001.

C. A. Zapffe, Hinrichs, Precursor of Mendeleev, *Isis*, *60*, 461–476, 1969.

R. Zingales, From Masurium to Trinacrium: The Troubled Story of Element 43, *Journal of Chemical Education*, *82*, 221–227, 2005.

R. Zingales, The History of Element 43—Technetium (the author replies), *Journal of Chemical Education*, *83*, 213–213, 2006.

AUTHOR INDEX

INDEX